CW01678230

Woody species in auxiliary roles

Arnoud Budelman

Woody species in auxiliary roles

Live stakes in yam cultivation

DEVELOPMENT ORIENTED RESEARCH IN AGRICULTURE

ROYAL TROPICAL INSTITUTE - THE NETHERLANDS

This study appears as volume 3 in the series *Development Oriented Research in Agriculture.*

CIP-DATA KONINKLIJKE BIBLIOTHEEK, DEN HAAG

Woody species in auxiliary roles : live stakes in yam cultivation / Arnoud Budelman. -
Amsterdam : Royal Tropical Institute. - (Development oriented research in
agriculture : 3)
Also publ. as thesis, Wageningen, 1991. - With ref.
ISBN 90-6832-041-6
NUGI 825
Subject heading: agriculture : Africa.

Cover design: Nederlands Ontwerp bv
Series editor: Rosemary W. Gunn
Printer: ICG Printing, Dordrecht

ISBN 90 6832 041 6
NUGI 825

Let's celebrate diversity!

Harold F. Miller

Acknowledgements

First, I would like to thank Prof. M. Flach and Prof. R.A.A. Oldeman for their long-standing critical support and encouragement. Many others, over the years, have also helped to shape my thinking regarding the subject of this book. Specific appreciation goes to Dr. P. Ahn, Dr. J. Beer, Prof. J.D. Ferwerda, Dr. P.A. Huxley, Dr. B.H. Janssen, Prof. I.C. Onwueme, Prof. C. Stigter, Dr. W.A. Stoop, Dr. J. Sumberg, Dr. W. Teel, Prof. M. Wessel, Dr. E. Westphal and Dr. R. Wijewardene.

I would like to convey my gratitude to the Editors of the Journal *Agroforestry Systems* for their kind permission to republish the papers, in particular Prof. H.J. von Maydell. Lastly I wish to thank the KIT Press staff at the Royal Tropical Institute, for their careful preparation of the book.

Table of contents

Preface

Yam production is much more problematic in the humid zones of West Africa than in the savanna. In humid areas, staking is essential if reasonable yields are to be obtained. But staking is very labour intensive, and accounts for a sizeable proportion of the cost of producing the crop. In recent years, as the forested area from which stakes can be obtained has grown progressively smaller, the costs of stakes and staking has continued to increase.

A second major problem for yam production in humid regions is the rapid depletion of soil nutrients, and, in some cases, the degradation of the physical condition of the soil. Indirectly, this has led to the use of shifting cultivation as a stable farming system. However, in recent times, land pressure has forced many farmers who traditionally practised shifting cultivation to adopt some measure of sedentary agriculture, which demands management strategies in order to maintain soil fertility.

The series of papers in this volume very comprehensively tackles the twin problems of stake scarcity and soil depletion in yam production in the humid region of West Africa. The papers go well beyond the usual platitudes about alley-cropping, environmental preservation and stabilization of the cropping system. They also quite critically analyse the reasons behind the advantages to be derived from live staking of yams. Although the author's analysis does not highlight this sufficiently, the physical and chemical contributions of the leaf litter (mulch) are certainly paramount.

Gliricidia sepium is the species of choice here, but presumably the choice of a species best for other environments would require a similar process of selection and experimentation. An important factor contributing to the suitability of a given species would be its nitrogen-fixing capability in the environment under consideration.

Certainly a species whose roots compete minimally with the target crop (an aspect explored by the author) and whose roots fix as much nitrogen as possible, would have much to recommend it. Several other criteria must of course be satisfied, and most of these are discussed. The ultimate practical test is, of course, the improved quantity and quality of yield of the target crop, and the stability of such improvement from year to year.

I seize this opportunity to congratulate the author and his collaborators and supervisors for a job well done. A sound scientific foundation has been laid for further work in this area, which has for many years been rather opaque. It is my hope that this work will lead to a cheaper, more stable, more sustainable system for yam production in the humid tropics.

Prof. I.C. Onwueme
Federal University of Technology
Owerri, Nigeria

I Woody species in auxiliary roles: live stakes in yam cultivation

1
A review

Staking is important in the cultivation of the climbing food crop yam, because it helps to achieve economically viable production levels. In many parts of West Africa, where yam is a main staple, stakes are difficult to obtain nowadays, due to deforestation. One solution is to grow yams together with suitable perennials. These perennials play an 'auxiliary' role, by acting as live stakes. The auxiliary species may also help to maintain the levels of organic matter and nutrients required in the soil for successful, sustained yam production, particularly if the auxiliary species chosen fixes nitrogen.

The decision to study live stakes in yam cultivation scientifically was inspired first by an actual situation involving yam cultivators faced with land shortages in southwestern Ivory Coast (Budelman and Zander, 1990), and second by reports of the use of live stakes by farmers in deforested areas of southwestern Nigeria (Clayton, 1958; Wilson and Akapa, 1981; Agboola et al., 1982). The underlying assumption is that a technology developed by farmers as a reaction to land shortages in one region may stand a good chance of being adopted in similar situations elsewhere.

In general terms, the objective of the study has been to contribute to the development of a low external input food cropping system that is productive and sustainable. Auxiliary crop species only indirectly contribute to the primary object of a cropping system – here, production of yam tubers. Sub-sets of research questions can be asked concerning both the products and services of the auxiliary crop species, such as the use of the leaf mass produced by the perennials as a mulch, helpful in environmental management of cropping systems; and, more specific to the cropping system under discussion, the use of the woody species as live stakes to support the yam crop.

On-station fieldwork took place in sub-humid Ivory Coast, West Africa, from October 1983 to December 1987. This research was reported in a series of papers published in *Agroforestry Systems*, which are reprinted here. They discuss the use of three nitrogen-fixing, woody perennials – *Leucaena leucocephala*, *Gliricidia sepium* and *Flemingia macrophylla* – as auxiliary crop species in the cultivation of the climbing, tuberculous food crop water yam, *Dioscorea alata*. These papers are arranged in three subject groups:

1. productivity and composition of the leaf material of the woody species: a quantitative and qualitative inventory of leaf material, considered as a

resource with respect to crop production;
2. use of leaf mulch in environmental management of the cropping system: weed control, soil fertility improvement, soil temperature and soil humidity control;
3. use of the woody species as an integral part of an annual crop production system.

The context

The practice of cultivating tree species in association with short-lived agricultural crops is probably as old as the history of agriculture. If one accepts that early man must have gained an intimate knowledge of plant species by collecting forest products, saving useful woody species while opening the forest can be seen as a logical step in the evolution towards cultivation. An example of a traditional farming system in which trees conserved from the original vegetation play a multi-purpose role is the Hanunoo case from the Philippines (Conklin, 1957). Tree-saving practices have also been documented in other case studies of traditional shifting cultivation systems in the wet tropics (Miracle, 1967; Clarke, 1976; Wilken, 1977; FAO, 1985; de Rouw, 1987 and 1991).

In all likelihood, crops like cinnamon, coffee, tea, cacao, vanilla, black pepper and yam, with origins in the forest ecosystems of the wetter tropical zones, were domesticated together with tree species that provided the shade and/or support necessary for sustained cultivation (for example, the common Spanish name for *Gliricidia sepium* is 'Madre de cacao', or mother of cacao; see further Conklin, 1957, p. 62; Budowski et al., 1984; Hladik et al., 1984).

The drier regions of the tropics, e.g. the Sudan zone in West Africa, provide even clearer examples of farmers' tree-saving practices. Indigenous trees such as *Faidherbia albida* (syn. *Acacia albida*; Anonymous, 1988), as a source of food for cattle, and *Vitellaria paradoxa* (syn. *Butyrospermum paradoxum*; Shea nut, Karité) are instrumental in the provision of edible fats (Seignobos, 1982) and are well-represented in fields carrying crops of groundnut, millet or cotton. In the case of *V. paradoxa*, the most productive trees are located in the cultivated areas, not in the surrounding bushland (Ruyssen, 1957); whereas it has repeatedly been suggested that *Faidherbia albida* increases the carrying capacity of the land (Charreau and Vidal, 1965; Pelissier, 1966; Lemaître, in Felker, 1978).[1] These two trees exemplify a large group of tree species that play a role in local farming systems (cf. von Maydell, 1983).

The economic rationale for practising conservation of trees is obvious: tree and shrub species provide a broad range of products. In addition to food, beverages and animal feed, woody species are sources of fibre, fish poison, construction and fuel wood, nectar for bees, medicaments (for both people and domesticated animals), dye, gum, soap, latex, material for thatching, and so forth (cf. Lévi-Strauss, 1950; Verdcourt and Trump, 1969; von Maydell, 1983).

Promoting trees is also rational from an ecological point of view; it conserves both land and genetic resources. Farmers often[2] have profound knowledge of the phenology of woody species (cf. Felker, 1978, pp. 108–120), and deliberately create favourable conditions for site regeneration (Wiersum et al., 1985). Among the conservation practices observed are:

- selective weeding, sparing seedlings of valuable tree species;
- protection of sprouts from tree stumps;
- minimum tillage;
- planting or transplanting of trees (Miracle, 1967, pp. 108-109; Manner, 1981).

In looking at the role trees play in subsistence farming, it is often hard to draw a line between cultigens and useful wild species. Such an integrated perception of the role of trees is not common outside some of the traditional farming systems mentioned. Until late in the 1970s science – and even more, policymaking bodies – maintained a distinction between matters concerning trees, or forest in general, and those dealing with agriculture (King, 1979). In fact, this situation still exists in most countries: forestry and agriculture are housed in separate ministries or departments, which naturally creates barriers to communication and cooperation. Although this is not the place to trace the origins of the separation, brief comments on some recent developments will help to put the contents of this volume into a perspective.

The book *Trees, food and people* by Bene and co-authors was a conceptual breakthrough in the scientific community. Explicit attention was given to the integration of agriculture, forestry and, if need be animal husbandry (Bene et al., 1977). This report, plus institutional consultation and international fund raising resulted in creation of the International Council for Research in Agroforestry (ICRAF) (King, 1987). In the following decade, agroforestry has become accepted in the scientific community as a valuable land management strategy (Steppler and Nair, 1987).

As this historical note makes clear, as the object of scientific investigation and policymaking, the integrated agroforestry approach to land use is in its infancy. Nevertheless expectations are high. Over the past decade agroforestry has been promoted, not without a sense of urgency, as an alternative to shifting cultivation. In that context we must deal with considerable problems, since providing for permanent rainfed cultivation of annual crops, in a balanced system of continuously high productivity in both humid and semi-humid areas, remains technically difficult (Ruthenberg, 1976). The most important constraints are low natural soil fertility and poorly developed infrastructure (markets, roads, agricultural services).

While integrating trees may help to improve sustainability of annual crop production systems, a natural paucity of plant nutrients cannot be cured by the mere presence of woody perennials. The tropical Oxisols and Ultisols[3] are notoriously poor in nutrients, while the little chemical fertility there is tends to be labile. Within months after burning of the standing vegetation, significant changes may occur in the extent to which the cation exchange capacity (CEC) is saturated, due to leaching (Sanchez, 1977; Seubert et al., 1977; Budelman and Zander, 1990). Moreover, the CEC often depends heavily on the percentage of organic matter in the soil and can be damaged irreversibly when soils are mismanaged. The length of the fallow period needed to restore chemical and physical soil conditions to levels comparable to the original situation is an indication of the damage done to the ecosystem. After an agricultural production cycle, more than 15 years is often needed for recovery.

Whether agroforestry as a strategy of land management, and as an alternative to the traditional long fallow system, will provide an answer to the problem of sustained annual cropping in fragile tropical ecosystems is still an unsolved question. Both Young (1987, p. 139) and Sanchez (1987) have cast doubt on this possibility. The latter author observed that the best documented successes of agroforestry systems are generally located on relatively fertile soils, and that much more, and more rigorous, experimental testing is required to prove that agroforestry systems really improve physical properties of the soil, maintain soil organic matter, and efficiently recycle nutrients.

Secondly, successful agroforestry system development hinges on generating more information concerning its components. Compared to knowledge about crops, research on auxiliary tree species is in a primitive state (Steppler, 1987). Obviously the same holds for data collection at the level of crop/auxiliary tree systems.

There are, however, literally hundreds of agroforestry systems in the tropics, and equal numbers of 'new' systems can potentially be developed by reassembling and adapting these to local needs and circumstances. Thus in generating new systems, a pragmatic approach and clear cut objectives are essential. In that respect Steppler's challenge, to seek '...a specific ideotype of multipurpose tree to fit a particular niche' (Steppler, 1987, p. 17) has been most helpful in defining the boundaries of this study. The niche chosen here includes the requirements of a climbing food crop; the purposes are to provide live stakes and mulch to be used in management of the cropping system environment.

Live stakes in yam cultivation: a case study

The eight publications[4] introduced in this Chapter make up a case study in which three leguminous perennials – *Leucaena leucocephala, Gliricidia sepium* and *Flemingia macrophylla* – were tested as auxiliary crop species in yam cultivation.[5]

In agroforestry literature, this kind of tree-based cropping system is classified as an 'agrisilvicultural system, aimed at zonal/associated intercropping of annual crops, and fallow improvement' (Nair, 1987). The primary aim of the field research has been to collect basic data (cf. von Carlowitz, 1986).

Criteria for selection of species. Firstly, a 'standard', a relatively well-researched species, is needed to put results in context. *Leucaena leucocephala*, well-known from alley-cropping experience all over the tropical world, plays that role here (cf. Kang et al., 1984).

Secondly, although *Gliricidia sepium* was known to have been used by farmers as a live support in yam cultivation (Clayton, 1958; Wilson and Akapa, 1981; Agboola et al., 1982)[6] little concrete information was available regarding the species' performance under such circumstances. However, the evidence that was available helped to formulate the specific objectives of the study and to choose the test crop, the climbing food crop yam.

Thirdly, *Flemingia macrophylla* was chosen because of two features that offer a contrast to the other two species: its leaves decompose relatively slowly; further, it is a shrub, and consequently has a different basic architecture. Such contrasts are interesting in view of the function required – an *in situ* grown support system. Very little documentation on this species exists (Budelman, 1989d).

Climate and soil. The trials reviewed here were conducted in the Ivory Coast near Abidjan (5°19'N, 4°13'W), on an experimental farm approximately at sea level. The climate is sub-equatorial, with two rainfall peaks. Table 1 gives a summary of climatic data. The soils of the experimental farm belong to a small band of tertiary and quarternary deposits along the

Table 1. Summary of climatic data at the trial site

	Temp. min. °C	Temp. max. °C	Air humidity min. %	PET* mm yr^{-1}	Hours sun-shine yr^{-1}	Global radiation $10^6 J m^{-2} yr^{-1}$	Average rainfall in mm yr^{-1}	Days with rain yr^{-1}
Average	22.4	30.0	70					
Total				1249	1787	5666	2131	130

* PET (potential evapo-transpiration), calculated according to Turc's formula
Source: Adiopodoumé, Ivory Coast. Adapted from Monteny and Eldin, 1977

Table 2. Selected physical and chemical parameters of the soil of the trial area (0–20 cm)

% sand 50-2000 µm	% clay + loam < 50 µm	pH H_2O	O.M. %	C ‰	N ‰	P* ‰	CEC† Meq.	Base saturation %
84.5	14.1	4.7	1.06	6.11	0.59	0.24	3.27	13.6

* total $P(HNO_3)$
† per 100 g soil
Source: abbreviated following Siemonsma, 1982

coastline. The dominant characteristic of these sandy, ferralitic soils is their strong desaturation, combined with a weak base exchange capacity (Table 2). Potassium is the most limiting of the major elements (Roose and Chéroux, 1966).

Data summary. For descriptions of the trial designs and other methodological aspects, the reader is referred to the methodological notes in the chapters that follow. Results of the study are summarized in Table 3. The information is presented relative to an index. For example, when providing information on the nitrogen content of the leaves, the species with the highest content is assigned an index value of 100, while the nutrient contents of other species are presented as a percentage of this value. The column at the far right gives absolute figures for the species assigned the index value of 100. Where relative quantification is impossible, differences are presented in descriptive terms.

The study demonstrates that each of the three auxiliary crop species has a specific 'service-profile'. Table 3 shows that *Leucaena leucocephala* excels in providing great quantities of nutrient-rich, easily decomposable leaf matter (Table 3, items 1, 2, 3, 4, 9), but, at the same time, is a strong competitor to neighbouring plants (items 10, 13), may exhaust soil fertility when the biomass produced is not carefully used (Chapter 3), and has only limited value in soil microclimate conditioning and weed control when leaf material is used as mulch (Table 3, items 5, 6, 7).

Leucaena leucocephala, because of its robust performance, is a typical example of an auxiliary crop species that should be used in a zonal planting configuration (cf. Huxley, 1985), such as alley-cropping. When planted too close, the main crop may show a reduction in yield (Chapter 8).

In comparison to *Leucaena leucocephala*, *Gliricidia sepium* is much less aggressive: witness its combination of lower leaf mass productivity with lower root density in the upper soil stratum (Table 3, items 1, 10). This in part explains its success in crop production systems that require close association of auxiliary crop species and main crop, as is the case when the perennial is used as live support for a climbing crop (Table 3, item 13; cf. Huxley, 1985; Chapters 8 and 9).

The potential importance of *Flemingia macrophylla* lies in the composition, size and behaviour of its leaves while they are drying. A relatively high C/N ratio partly explains the slow decomposition of the leaves and the ongoing effect in terms of suppression of weed seedlings, lower soil temperature and higher soil moisture content in comparision to the other two mulch types (Chapters 4, 5 and 6).

The yam/live stake interface: potential and problems

Of the three leguminous species tested as live support systems for yams, *Gliricidia sepium* proved most suitable (Chapter 8). Table 4 sums up the results of the detailed study on the *Gliricidia sepium*/yam association presented in Chapter 9. In this Table, mulch application and live stakes, as well as the combination of these two factors, are taken as different treatments; their effects on early development, leaf nutrient status and tuber yield of the yam crop are shown. The Table illustrates both the potential and problems of tree-based annual cropping systems. The combined effects of mulch application, light shade and live stakes on early crop development and tuber yield is significant. Compare for example in the Table the results of treatments 'a' (no mulch, no stakes) and 'd' (mulch and stakes, both produced *in situ*). Mulching alone (treatment 'b') also significantly improves the yield of the crop (see also Chapter 7).

Treatments 'b' and 'd' differ in that experimental units of treatment 'b' received mulch produced elsewhere (namely from the experimental units under treatment 'c'), while the mulch used in treatment 'd' was produced

Table 3. Auxiliary crop species characteristics: a relative appraisal

Item	*Leucaena leucocephala*	*Gliricidia sepium*	*Flemingia macrophylla*	Index value 100
1. *Capacity for leaf dry matter productivity* [a] (100 = highest)				
- at 1,000 pl ha-1	100	89	67	5.7t y^{-1}
- at 15,000 pl ha^{-1}	100	66	81	17.2t y^{-1}
2. *Nutrient composition* (100 = highest)				
- nitrogen concentration	100	94	78	3.37%
- phosphorus concentration	92	100	92	0.24%
- potassium concentration	90	100	68	1.68%
- calcium concentration	69	100	48	1.34%
- magnesium concentration	71	100	42	0.48%
3. *Size of leaflets* [b]	very small	small	large	
4. *Decomposition of leaf matter* [c] (100 = most persistent, most digestible)				
- persistence	57	41	100	53.4 days
- *in vitro* digestibility	100	99	48	
5. *Effect of leaf mulch on retarding weed emergence* [d]	no effect	limited effect	most effective	
6. *Effect of leaf mulches on soil temp. reduction* [e] (at 5 cm depth in soil, initial quantity 5 ton ha^{-1} DM)				
- initial impact, after 10 days (highest impact = 100)	57	76	100	-9.8°C
- average impact, over 60 days (highest impact = 100)	44	70	100	-6.6°C
- effective lifetime of the mulch material	shortest	intermed.	longest	

on the spot. Experimental plots of treatment 'b' therefore experienced an addition of plant nutrients from an outside source, while in treatment 'd' nutrients were made available via the leaf mulch produced by the *Gliricidia sepium* trees grown *in situ*.

Whether the higher potassium content of yam leaves in treatment 'd' actually comes from nutrients extracted by the trees from soil layers that cannot be reached by the yam roots (a process referred to as 'recycling') cannot be ascertained from the present data. What is demonstrated, however, is that the trees act as 'concentrators' of this scarce nutrient. With proper management, e.g. regularly putting the potassium-rich leaf mass produced by the auxiliary trees close to the yam roots, the nutrient flush

Table 3, continued

Item	*Leucaena leucocephala*	*Gliricidia sepium*	*Flemingia macrophylla*	Index value 100
7. *Effect of mulch on soil moisture conservation*[e]				
(0-5 cm, initial quantity 5 ton ha^{-1} DM)				
- initial impact, after 10 days (highest impact = 100)	71	91	100	+5.6%
- average impact, over 60 days (highest impact = 100)	50	85	100	+4.6%
- effective lifetime of the mulch material	shortest	intermed.	longest	
8. *Nyctitropic movements of leaflets*	yes, strong	yes	yes	
9. *Coppicibility and rate of regrowth*	excellent	good	good	
10. *Fine roots*[f] *in upper 40 cm soil profile* ($g\ m^{-2}$; 100 = highest)				
- at 1,200 pl ha^{-1}	100	73	39	67 g
- at 8,000 pl ha^{-1}	100	48	57	126 g
11. *Propagation*	seed	seed cuttings	seed	
12. *Risk that species turns into weed*	high[g]	very low	low	
13. *Suitability as live stake in yam cultivation*[h]	low	high	low	

a) based on 3 month regrowth cycles, Chapter 2

b) *L. leucocephala*, 0.7-1.0 cm; *G. sepium*, ca. 5 cm; *F. macrophylla*, 10-20 cm

c) persistence based on 'half-life' values for the mulch material, Chapter 4

d) seedlings only, initial quantity 4 ton ha^{-1} DM, Chapter 5

e) Chapter 6

f) roots < 1 mm, Chapter 8

g) dependent upon cultivar used, see Brewbaker (1987)

h) Chapters 8 and 9

from the decomposing leaf matter significantly improves tuber yields (Chapters 7 and 9). In other words, the nutrient-use efficiency of a tree-based yam production system is higher than the efficiency of a system in which yam is solecropped.

That same efficiency, however, is reason for concern. Species such as *Leucaena leucocephala* and *Gliricidia sepium* extract considerable quantities of nutrients per unit of surface (Chapter 3). When the leaf mass is regularly cut and used elsewhere,[7] chemically poor soils are likely to be rapidly exhausted.

The ability of tree-based systems to exhaust soils is demonstrated in Table 4, treatment 'c', where the mulch produced by the trees is removed

Table 4. Data summary: the *Gliricidia sepium*–yam association

Treatment	a	b	c	d	
	no mulch	+ mulch*	no mulch	+ mulch*	
	no stakes	no stakes	+ stakes	+ stakes	
yam plants ha^{-1}	20,000	20,000	13,300	13,000	
live stakes ha^{-1}			3,300	3,300	
nutrient transport via mulch		addition, from 'c'	extraction, to 'b'	neutral, recycling	L.S.D. (P=0.05)
Early development					
plants sprouted §	42.5 [a]	62.7 [a b]	68.2 [b]	83.0 [b]	24.2
sprout length in cm	21.6 [a]	31.3 [b c]	26.8 [a b]	36.6 [c]	8.3
soil cover %	36.0 [a b]	44.4 [b]	25.6 [a]	61.6 [c]	16.4
stakes occupied by yam vines†			9.0 [a]	18.0 [b]	5.4
Nutrient content of yam leaves					
nitrogen %	3.13 [a]	3.23 [a]	3.01 [a]	3.22 [a]	N.S.‡
potassium %	1.02 [a]	1.93 [b]	0.80 [a]	1.76 [b]	0.68
Tuber yield					
kg per yam plant	0.44 [a]	0.89 [b]	0.54 [a]	1.55 [c]	0.29
tons per ha	8.8 [a]	17.8 [b]	7.2 [a]	20.7 [b]	5.1

* during the cropping cycle, a total of 4.8 ton ha^{-1} DM of leaf mulch material was applied in both treatments 'b' and 'd'. The mulch was distributed at *equal rates* of DM/unit of surface

§ results with the same superscript, and on the same line, do not differ significantly

† out of 26 live stakes sampled per experimental unit

‡ not significant

Source: Budelman, 1990b

from the experimental units and used as green manure for the units of treatment 'b'. The effect of this practice is visible, although not statistically significant, in the low potassium content of the yam leaves in treatment 'c'.

A problem specific to the use of organic fertilizers to improve crop production is that organic mulches bear no label telling which nutrients they make available and in what quantities. Contents vary with the species considered, depend on the age of the leaf material, change over time, and depend on soil conditions in the location where the mulch was produced. Mulch decomposition depends not only on the chemical composition (C/N and C/P ratios, for example) of the material, but also on the temperature and moisture regimes during decomposition. Moreover, nutrients are released from leaf mulches at different rates (Chapter 4).

Thus there is clearly little scope for specific recommendations aimed at optimizing the use of mulch in cropping system management. Generally, however, the logical time slot for mulching annual crops would be early in their cultivation cycle. At that particular stage in the annual crop cycle, mulches are valuable as soil cover, and help to enhance the acceptance of rainwater. Use at this time also assures maximum benefit in terms of water conservation and lower soil temperatures; moreover, the crop will profit at a later stage from the nutrients released (Chapter 4).

Uncertainty about the behaviour of mulches is, however, only part of the 'black box', experimentally speaking, that tree-based annual cropping systems represent. When screening woody species with respect to their suitability as live stakes for climbing crops, the obvious criterion is the yield of the main crop. To explain differences in crop yield level as a function of the auxiliary crop species used, it is logical to look at the auxiliary crop species' production potential, its root density in the zone where the main crop roots, plus less easily measurable aspects such as the architecture of the perennial species, and the light conditions under its canopy (Chapter 8).

However, quantification of these characteristics provides at best circumstantial evidence. For example, it is the actual competition from the roots of the woody species that counts, not the observed root density as such. Even with a relatively low root density, an auxiliary crop species may be a fierce competitor, for example due to allelopathic qualities. The problem, of course, is that the actual root competition between the species is hard to measure in field trials.

Similarly, results of field experiments that combine woody auxiliary crop species with annual crops are difficult to interpret, because effects are so thoroughly confounded. The data in Table 4 do not unequivocally demonstrate which factor primarily determines the yield increment that results when *Gliricidia sepium* is integrated in yam cultivation (Chapter 9): is it due to the live stakes, or to the mulch produced *in situ*?

Including dead stakes (e.g. made from bamboo) as part of the treatment series would not have solved this particular problem. In an analysis of variance, the effect of staking on the yield cannot be separated from the effect of the mulch generated by live stakes. If one compares dead stakes and live stakes according to their respective effects on yam yields, and chooses to remove the leaf mass from the live stakes, extraction of nutrients will occur. This puts the live stake treatment at a disadvantage. As Table 4 shows, such extraction may lower crop production. Conversely, if in the live stakes treatment the leaf mass is left as mulch for the experimental units concerned, those experimental units including yam plants growing on dead stakes will be treated unfairly.

Such problems in interpretation of results confirm Steppler's observation that agroforestry systems research requires different experimental designs and statistical techniques from those developed for solecropping (Steppler, 1987, p. 16).[8] The question is, however, the extent to which interactions must be explained and quantified before the general validity of agroforestry techniques is accepted. After all, the nature of the subject is such that no single component of a tree-based cropping system is likely to offer dramatic potential for improved output. Consequently, on-station research aimed at optimizing and fine-tuning the tree-based annual cropping model makes little sense.[9]

The effect of mulching, for example, is derived from the sum of many marginal improvements in the cropping system environment: a slightly lower soil temperature, a bit more water caught, weed seeds that do not germinate due to lack of light, a slightly higher nutrient content in the topsoil, a few extra root nodules where *Rhizobium* bacteria fix nitrogen, and so forth. Thus mulch-induced effects work together to stimulate crop seed germination and early seedling growth, resulting in a crop that closes earlier (see Table 4, 'early development': 'soil cover', and Chapter 7).

Auxiliary crop species in annual crop systems: considerations in selection

Auxiliary crop species contribute only indirectly to the economic output of a crop production system. When auxiliary perennials are interplanted, their intended role is to protect and stabilize the cropping environment, by limiting erosion and/or leaching, reducing temperature fluctuations, and so forth. Auxiliary crop species may, however, produce secondary products, such as an occasional batch of fodder or firewood. Such by-products increase the economic attractiveness of a tree-based annual cropping system.

Studies explicitly intended to define ideotypes of auxiliary crop species suited to a particular cropping system niche are rare (Steppler, 1987; Wayne S. Teel, 1990, personal communication). Beer (1987) lists desirable characteristics for auxiliary (shade) trees to be used with commercial tree crops (coffee, cacao, tea) but to date there has been no such review with respect to woody species in annual cropping systems.[10] This is understandable, given the enormous range of possibilities: the potential number of combinations of auxiliary crop species and annual crops, variability in climatic and soil conditions, and variation in farmers' objectives, backgrounds and management skills.

The results of the present study do not permit a general listing of the properties of auxiliary crop species with respect to the design of tree-based annual cropping system. The number of auxiliary crop species tested is too limited for that purpose, and the range of functions studied is too narrow.

To focus the discussion on selection of auxiliary crop species, the 'case' of *Leucaena leucocephala* will be briefly considered. This perennial was at one time almost the sole standard in the budding alley-cropping technology, and its history and performance teach some important lessons. This woody legume was initially widely extolled as a 'wonder tree', a 'saviour' (Benge, 1981), qualifications that stimulated an often unwarranted introduction in the tropical world. Two important properties account for its popularity: high biomass production potential and an ability to fix uncommonly large quantities of nitrogen.

Singleminded distribution of *Leucaena leucocephala* has, however, repeatedly produced disappointing results, due to unfavourable climatic or soil conditions. Although 1,200 mm annual precipitation must be seen as the lower limit for the rainfall requirements of the species, rainfed Leucaena planting has occurred in regions receiving annual rainfall of 900 mm or even less. Soil acidity, a problem frequently encountered in tropical Oxisols, affects seedling establishment and early growth, as well as later biomass productivity. In some cases, its prolific seed production has turned the species into a serious weed, which is difficult to eradicate manually.

The narrow genetic base of the germ plasm distributed started to cause problems in late 1982, when the Leucaena Psyllid (*Heteropsylla cubana* Crawford) emerged as a major pest outside its native range in the Americas (NFTA, 1988). Increasingly, questions about the species' ability to check soil erosion have appeared (Halos, 1980; IDRC, 1982, p. 173). Its fast leaf litter decomposition (Chapter 4), capacity to shade out undergrowth, and nyctitropic leaflet movement during heavy rains, make it likely that Leucaena, when planted singly, will in fact be an erosion hazard.

Finally, a decade of alley-cropping research in tropical Africa, primarily involving *Leucaena leucocephala* as auxiliary crop species, has received very little response from farming communities (Sumberg and Okali, 1988), even though auxiliary tree use and planted fallows are not uncommon in local farming systems (Grove, 1951, p. 57; Clayton, 1958; Wilson and Akapa, 1981; Agboola et al., 1982; Raintree and Warner, 1986; Adejuwon and Adesina, 1990).

This list of observations is not meant to discourage the use of *Leucaena leucocephala* as an auxiliary crop species. The thesis, however, is that many of these problems would not have occurred, or would have been less serious, had policymakers in agricultural research and rural development opted for a broader base of candidate species. Overuse of *Leucaena leucocephala* appears to stem in part from lack of knowledge of alternative species.

The observations discussed suggest that the most compelling rule in auxiliary crop species selection is to exploit biological diversity, rather than betting on a few high-performing species or varieties thereof. This conclusion implies that reliable data on key characteristics must become available for a range of promising species.

In selecting auxiliary crop species, two categories of criteria can be distinguished. The first includes several key criteria, generally considered to be of primary importance:[11]

- biomass productivity;
- resilience to frequent lopping, with fast regrowth;
- leaf nutrient content and nitrogen fixation;
- plant architecture, including rooting patterns.

The second category includes situational criteria: properties of auxiliary crop species that respond to particular environmental constraints, or requirements of main crops in the cropping system under consideration. In this latter category, for example, the sharp-leaved *Agave sisalana* (sisal) and unpalatable *Euphorbia* and *Jatropha spp.* are found; they are suitable under conditions in which freely-browsing cattle inhibit the establishment of other species.

Properties of perennials specifically required by a main crop are also in this category. Live stake species used in pepper cultivation, for example, must have a rough bark that does not come off easily (cf. Marinet, 1955). Further, the wood of live stake species must be sufficiently strong not to collapse under the weight of the leaf mass of a given climbing crop (Chapter 8).

Biomass productivity. When resource production for agricultural purposes is the primary aim, biomass productivity of an auxiliary crop species will be a key parameter in cropping system design. However, it should be borne in mind that highly productive auxiliary crop species, such as *Leucaena leucocephala*, *Cassia siamea* and others, are at the same time important competitors to a main crop. Integrated in a cropping system, they make use of the same limited resources – light, water and nutrients – that feed the main crop or crops.

As a rule, integration of strong producers demands zonally arranged cropping systems, such as alley-cropping (Huxley, 1985). When close association of main crop and auxiliary crop species is required, as when the latter is used as a live support, biomass productivity levels of the perennial species will necessarily be lower (Chapters 2 and 8). Obviously, close association of auxiliary crop species that are too competitive to main crops will have to be avoided. Zonal planting arrangements, however, may stem from requirements other than the properties of the cropping system components. Mechanization of farm operations, e.g. ploughing, demands zonal planting. In making sustainable use of sloping land, such a cropping system organization is instrumental in terrace formation, and hence in erosion control (Roose, 1986; Garrity, 1989).

Resilience to lopping. Fast regrowth after pruning and the ability to withstand frequent lopping are prerequisites for selection when biomass production for environmental management of the cropping system is the main reason for introducing an auxiliary crop species. Regrowth depends on the availability of active meristem tissues all over the plant, plus a sufficiently flexible growth model. A combination of these properties is often found in Leguminosae (Oldeman, 1986), but is also seen in plant families such as Euphorbiaceae, Compositae, Rosaceae and others.

Nutrient content. Nitrogen fixation by bacteria in root nodules is often seen as one of the most important assets of auxiliary crop species. Leaf nitrogen content in species that fix nitrogen is higher than in species that do not fix this essential plant nutrient.

The ability to fix nitrogen is widespread in the Leguminosae family (Allen and Allen, 1981); hence the disproportionate attention to leguminous species in early alley-cropping research and literature (e.g. Kang et al., 1984, p. 4). In fact, apart from a few exceptions (*Leucaena leucocephala* and *Casuarina equisetifolia*), actual fixation rates are unknown (Dommergues, 1987); high rates appear to be the exception rather than the rule.

The relationship between nitrogen fixation and improved soil fertility is far from conclusive. Felker (1978, pp. 88-92), in reviewing the literature on *Faidherbia albida* (syn. *Acacia albida*), concludes that the higher overall soil fertility found in the neighbourhood of this particular tree must be the result of its ability to fix nitrogen, but Sanchez (1987) notes that Felker's review does not contain a single study in which soil fertility was measured before and after the planting of *F. albida*. Thus the question of whether the tree's natural distribution is not biased towards more fertile sites must be raised.

Generally, if tree-generated soil enrichment takes place, the nutrient

Table 5. C/N ratio, productivity and disappearance of leaf material of three auxiliary crop species

Species	C/N ratio	Leaf productivity kg ha^{-1} day^{-1} DM at 5,000 plants/ha*	disappearance rate kg DM day^{-1}†
L. leucocephala	12.1	34	65
G. sepium	12.1	24	91
F. macrophylla	21.0	27	37

* based on productivity trials, Chapter 2

† disappearance rates are based upon half-life values for carbon, which makes up approximately 50% of the leaf DM. Mulch quantity in decomposition trials initially amounted to 4,000 kg DM ha^{-1}, Chapter 4

influx, aside from nitrogen, will be small; consequently, the process of enrichment is a slow affair (Kellman, 1979). Moreover, there is evidence that nutrients are concentrated around trees at the expense of adjacent treeless sites (see e.g. Belsky et al., 1989; Kessler and Breman, 1991).

Recently, authors have stressed the fact that the traditional research emphasis on nitrogen cycling may have been exaggerated (Huxley, 1986; Beer, 1988). They argue that in many situations nutrients other than nitrogen limit crop production. In such cases, perennials that manage to concentrate specific scarce nutrients are of interest.

More importantly, such studies advocate paying attention to species that combine high litter productivity with a slowly decomposing leaf material. This option facilitates nutrient loss prevention. Establishing litter layers made up of mulches with C/N ratios around 20 to 30, rather than 10, not only reduces removal of nutrients by surface runoff,[12] but also exploits the slow nutrient release property of these mulches. Leaf mulches with a C/N ratio of 10 (including the leaf material of *Leucaena leucocephala* and *Gliricidia sepium*) decompose rapidly, increasing the risk that crops will not be able to absorb the quantity of nutrients offered by the mulch (Chapter 4).

High nitrogen content and slow decomposition of leaf material are incompatible properties within a single plant species. Even when species manage to produce great quantities of leaf material per unit of time and space, as does *Leucaena leucocephala*, the low C/N ratio of the material prevents the establishment of a permanent litter layer. In such species, it takes more time to produce the leaf material than to lose it: see Table 5.

In selecting auxiliary crop species, nitrogen fixation does not automatically take priority. As noted, where nitrogen is not the primary constraint on plant production, and soil erosion is a serious hazard, cropping system sustainability is likely to be more greatly enhanced by planting auxiliary crop species with tough, slowly decomposing leaves, so as to maintain a permanent litter layer.

Architecture. Particularly under conditions of associated cropping, architectural aspects are important in selection of auxiliary crop species. Optimizing spatial complementarity of the species that make up the crop system is a primary goal in auxiliary crop species research (Willey, 1979). This search for complementarity concerns both aboveground and belowground characteristics of species. Leaf characteristics (e.g. size and colour) and quantity and dispersion of the leaf mass in the canopy determine the quality and quantity of the light transmitted (Willey, 1975; Budowski et al., 1984). Although hard to quantify, the structure of the canopy is one of the factors held responsible for variations in crop yield level observed when different auxiliary crop species are used (Chapter 8).

More specifically, in selecting auxiliary crop species for use as live supports, an elongated, unbranched architecture is desirable. Consequently, perennials that naturally produce long, slender branches are of interest (Chapter 8). Marinet (1955, pp. 330-332), for example, describes how *Albizia sp.* saplings are manipulated into becoming proper live stakes for black pepper, decreasing the hazard of root competition and encouraging a proper stature. Albizia's ability to propagate vegetatively is crucial to this procedure. Saplings are cut at a certain age, and then placed at a soil depth that will maintain sufficient distance between the rooting zone of the live stake and that of the associated pepper crop.

Vegetative propagation is important to other auxiliary functions as well. When auxiliary tree species are used to reclaim grass-infested areas, the ability to propagate vegetatively is fundamental to successful establishment. *Gliricidia sepium* has, in this respect, a clear advantage over seed propagated species such as *Leucaena leucocephala*. The former species has acquired some recognition in the battle against *Imperata cylindrica* (Clayton, 1958; Aken'ova and Atta-Krah, 1986).

Deep rooting is seen as a positive quality of auxiliary perennials. Both recycling of nutrients from deeper soil layers and less troublesome crop/auxiliary tree competition depend on this feature. Although root system development is influenced by local soil conditions, such as depth of the rootable zone, presence of hardpans and layers of concretions, nutrient distribution, water table, and so forth, there is also a species-specific element. Coster (1932) distinguishes three basic types in the root architecture of woody species:

1. deep penetration, few superficial roots;
2. deep penetration, many superficial roots;
3. shallow root system, variable number of superficial roots.

These categories are not fixed. A species may start as the second type and over time develop into type 1, particularly when nutrients are present in sufficient quantities in the deeper soil layers (cf. Coster, 1932, on *Leucaena leucocephala* in deep, relatively fertile soils).

For the species covered by this study, little quantitative information is available on root architecture. *Leucaena leucocephala* is said to root deeply. In plants that were 9–11 months old, planted at Buitenzorg, Java, 6% of the total root length was found more than 3 m below the soil surface (Keuchenius, 1927). Jonsson et al. (1988) show that the feeder roots of this species are also well-represented in topsoil, a finding that is confirmed by this study (Chapter 8).

In chemically poor soils, perennial root development usually takes place where the scarce nutrient resources are located. If nutrient availability is mostly confined to the upper soil horizon, root competition is likely to occur between perennial and superficially rooting annual crops.

Generally, in selecting auxiliary crop species for intercropping systems, neither short-lived woody pioneers nor species that belong to the latest stage of succession are unconditionally suitable. Suitability of the first category is limited, because of competitive fast growth and its roots' strong colonization of the upper soil horizon. The second category is impractical due to slow initial growth.

No perennial species, it seems, offers an ideal combination of fast growth, deep root penetration and limited rooting in the topsoil in all life cycle stages (Hairiah and van Noordwijk, 1986). In other words, there is no such thing as an 'ideal' auxiliary perennial: each species presents a 'mixed bag' of blessings plus less desirable properties.

Recommendations: a research agenda

Agroforestry involves agricultural production systems in which trees and annual crops are grown simultaneously on the same unit of land. Growing trees in this setting implies sacrificing part of the yield of the annual crops. However, product diversification and yield and income stability are arguments in favour of agroforestry as a land use strategy.

In this review and the papers that follow, emphasis is on the role auxiliary perennials can play in environmental protection. Here productive space is sacrificed to achieve sustainability of land use. Environmental protection as motive for planting perennial species demands a specific place in a classification of agroforestry systems. Use of the generic name *integrated fallow farming* to cover all annual crop production systems in which woody perennials form a buffer against environmental deterioration is proposed. In system classification, subdivision should then distinguish between zonally organized and associated planting configurations, as discussed.

Introduction of auxiliary perennials into annual cropping systems creates two fundamental problems. One is that the productive space left over must provide outputs that are sufficiently high to make the auxiliary perennial-based cropping system economically attractive. The other problem is that, for the farmer, introducing auxiliary perennials and the associated cultural practices creates a major management problem (Garrity, 1989).

The search for solutions to these two problems requires making use of two different environments. Increasing production efficiency typically requires on-station agricultural research, while the management question must be resolved in close cooperation with eventual users. Recommendations for future research will therefore be developed along these two lines.

Priorities for on-station research

Land use problems (e.g. soil erosion or chemical impoverishment) partly stem from ill-fitting agricultural production systems, from production procedures that fail to respond to specific environmental characteristics and limitations. Even where the attempt is made to compensate nutrient losses by applying fertilizers, rainfed annual cropping systems show varying degrees of lack of sustainability, due to topsoil losses resulting from surface erosion. This observation seems valid not only in the tropics, but in the climatologically much more benign temperate zones as well (de Wit, 1989). The primary motive for integrating auxiliary perennials in annual cropping systems, therefore, would be to increase the buffer capacity[13] of the crop system, so as to minimize soil and plant nutrient losses. The functions of an improved crop system buffer capacity are:

- maintenance of organic matter in the soil;
- increased efficiency in use of nutrients, including those provided by applying inorganic fertilizers;
- creation of barriers to soil transport.

Although these three functions are defined separately, the text that follows makes it obvious that they are strongly entwined.

Figure 1. The crop/auxiliary species biomass equilibrium

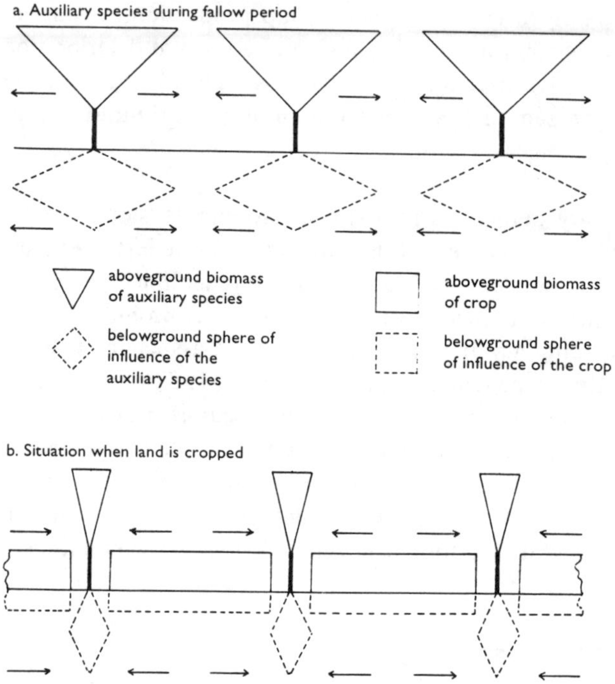

Maintenance of soil organic matter

Problems of soil degradation are especially notorious in soil types that depend on organic matter content to maintain soil fertility. In Oxisols and Ultisols, the activity of the available clay fraction is often low in terms of cation adsorption; 50% or more of the soil's cation exchange capacity – a major soil fertility indicator – may depend on the organic matter content of the soil (de Boissezon, 1970; Brams, 1971; Budelman and Zander, 1990). Under such conditions, soil fertility maintenance is almost identical with maintenance of the organic matter status of the soil (Sanchez, 1976).

If sustainable crop production depends on maintenance of organic matter in the soil, the question that must be answered is what portion of the land cultivated must be under auxiliary perennials, to provide enough organic material to make up for losses.

Young (1987) states that if the rest period needed to maintain productivity of the land is x per cent of time, the initial assumption must be that x per cent of the land must be occupied by auxiliary perennials. It is obvious, however, that the ratio between rest and productive time found in long fallow systems (e.g. one year of cultivation against seven years of rest: R=15%) is not feasible for an alley-cropping system. Although it is likely that modest fertilizer use will favourably alter the ratio between productive space and protective space (that occupied by auxiliary perennials), no data are available to suggest how the ratio would change given use of inorganic sources of plant nutrients.

Nutrient use efficiency

Auxiliary perennials act as a temporary store of plant nutrients. They play this role as long as their roots are alive, no matter whether the land carries a crop or lies fallow. If these live roots absorb nutrients from soil layers beyond the zone where the roots of the annual crops are active, auxiliary perennials potentially increase the availability of nutrients. When there are no crops in the field the nutrient storing effect of the auxiliary species is even more pronounced.

One can, therefore, define the auxiliary perennial/crop relationship as a dynamic equilibrium: the aboveground biomass of the auxiliary component is reduced as the crop develops, and left to expand as soon as the crop finishes its cycle (Figure 1). Cutting the biomass of the auxiliary crop species not only makes room for the developing crop, thus limiting interspecific competition; at the same time, the leaf part of the biomass in particular is a nutrient concentrate with a certain value as manure.

Putting the principles discussed here into practice requires, first of all, auxiliary perennials that root deeply and extensively. As argued earlier, information on rooting habits and development in auxiliary perennials is scarce, and therefore a priority for research.

Barriers to soil transport

Auxiliary perennials can contribute significantly to soil erosion control, when planted along contour lines and/or when species are selected to provide slowly decomposing mulch material. The standing biomass and living root systems of auxiliary crop species, as such, do not play major roles in preventing soil transport (Wiersum, 1985). However, when planted in a line and at close intervals, auxiliary perennials like *Leucaena leucocephala* effectively check soil erosion. This effect is enhanced when branches are placed along the line, close to the base of the trees.

These measures are known to assist in terrace formation (Roose, 1986), and therefore are potentially an important element in making hillside agriculture sustainable. Contour planting of auxiliary crop species is an accepted part of the so-called SALT, 'sloping agricultural land technology' (Garrity, 1989). A useful extension of SALT would be a search for woody species that are easier to establish than *Leucaena leucocephala*, and that produce notable quantities of slowly decomposing leaf material.

Organic mulch material carries nutrients that serve the interest of crop production, and prevents surface runoff and the resulting loss of nutrients. We need to ask what C/N ratio is optimal, in terms of providing these two beneficial aspects; or, would production of mixed mulches, using more than one perennial species (thus making available mulch material with both high and low C/N ratios) be a practical solution (cf. Siaw et al., 1991). Alternatively, would it make sense to opt for woody litter, leaving twigs and finer branches to rot, instead of removing these to facilitate agricultural operations (Huxley, 1986; Young, 1987)?

Conclusion

In crop production strategies designed for ecologically fragile areas, perennials may help to enhance the sustainable use of resources, but one cannot expect integration of woody species alone to guarantee sustained yields. Woody species do not generate fertility, at least not at the levels or in the timeframes required for intensive agricultural production.

Consequently, there is no doubt that inorganic fertilizers are needed, certainly under circumstances of low natural chemical soil fertility – at least to compensate for loss of nutrients exported in agricultural products leaving the field. Particularly in the case of continuous cultivation of rainfed annual crops, neither auxiliary crop species nor fertilizer application alone creates conditions that allow sustained production. A judicious mix of the two cultural practices may, however, be the key to successful crop production strategies with broad practical value.

The type of research questions described above demand on-station monitoring of long term trials, in which a sufficiently large number of woody auxiliary crop species is involved to allow effective exploitation of biological diversity. In addition to the creation of a solid data base, modelling is indispensable to the development of auxiliary perennial-based cropping systems. Since many of the relationships discussed in this review include elements of competition, in which the basic opposition is between production for the short term and protection of resources for the long term, optimization – finding the best possible balance between these two needs – appears to be at the heart of the matter.

Lastly, as this study demonstrates, questions regarding sustainability cannot be resolved unless research accommodates sufficiently long programme cycles. The reason the data provided by this study do not advance discussions of sustainability much further is essentially that the period of four years of field observations has been too short, a proof that experimental work with trees requires a different time horizon than is usual in research on annual crops.

On-farm research

The research priorities described reflect a belief that the agenda for on-station research should be limited. In addition to collecting data on characteristics and performance of the broadest possible set of auxiliary crop species, on-station research should focus on the biological underpinnings of the way auxiliary-based cropping systems function, and refrain from making precise operational prescriptions (Sumberg and Okali, 1988; Garrity, 1989). Thus the relative value of slowly decomposing mulches belongs to the domain to be studied, but not the relative spacing of a particular combination of crop and auxiliary perennial, nor the question of whether alleys should be three or four metres wide.

As this review makes clear, auxiliary perennial-based annual crop production technology, in spite of apparent ecological soundness, presents several serious drawbacks that stand in the way of rapid dissemination: complexity in terms of management, and the effort required for establishment. Implementation has consequences at all levels of a farmer's crop production system – labour allocation, tillage practices, soil fertility management, weed and disease control, and so forth – ramifications that as noted may well explain the limited enthusiasm of farmers for adopting, for example, alley-cropping.

Such questions can never be resolved unless the technology is put into practice, and farmers' handling of it closely monitored and evaluated. The hope is that in the process of on-farm research, farmers will adapt the technology to the hitherto unidentified constraints they meet (Sumberg and Okali, 1988). Further, it is to be expected that their experience will enrich the set of options, and that 'new' tree species, cultural practices, and so forth will be added. This perspective demonstrates the need for seeking flexibility in the technologies offered, rather than blueprint-type prescriptions.

Finally, on-farm research would gain enormously in terms of efficiency if farmers with solid experience in technology adaptation were to be mobilized to train other farmers (Garrity, 1989). Of the three stages in development of cropping system technologies – technology creation, testing and adapting, and transfer of information – institutionally speaking, it is this last component that represents the greatest challenge.

Notes

1. Some authors believe, however, that *F. albida* only establishes on relatively fertile spots. Higher fertility around these trees thus does not constitute proof that the species improves sites (cf. Sanchez, 1987)
2. Obviously, not all farmers display ecologically adapted cropping strategies. In many countries in Latin America colonists literally waste land (cf. Watters, 1971), while recently migrated farmers in Africa tend to maintain tillage techniques unsuitable to their new habitats (cf. Budelman and Zander, 1990)
3. Together they account for over 40% of potentially arable land resources in the tropics (Sanchez and Salinas, 1981).
4. References in the text are to Parts II–IV of the book; chapter numbers refer to specific reprints of papers (see Table of Contents).
5. Originally *Cassia siamea*, a non nitrogen-fixing woody legume, was also included in the live stake research programme. Trial plantations, however, suffered from a serious attack of an unidentified stem borer. This apparently caused a black coloration of the stems, and eventually led to the death of almost all plants. The species was subsequently eliminated from the programme. A farmer-based field trial including *Cassia siamea* as live stakes in the Taï region, southwest Ivory Coast, demonstrated that the species is not suitable in the role of a live stake (Budelman and Pinners, 1987).
6. See also Raintree and Warner, 1986; and Adejuwon and Adesina, 1990.
7. Cf. 'cut-and-carry' systems for livestock production (Francis and Atta-Krah, 1989).
8. See further Huxley, 1986; Huxley and Mead, 1988; Huxley et al., 1989.
9. Sumberg and Okali (1988) argue that only global proof of the validity of a new crop technology is needed, and that farmers should be involved as soon as possible in refining a technology, and adapting it to local constraints and needs. The experience of Francis and Atta-Krah (1989) seems to support this argument.
10. Some publications refer to characteristics of auxiliary species, albeit not systematically. For alley-cropping, see e.g. Kang et al., 1984, p. 4. Regarding live stakes in vanilla cultivation see e.g. Anonymous, 1970; Ashley, 1980. Live stakes in black pepper cultivation are covered by e.g. Marinet, 1955 and Seibert, 1987.
11. Over 50 specialists in agroforestry research were consulted and asked to indicate priority properties for the selection of auxiliary crop species.
12. Wiersum (1985) has shown that, in a plantation of *Acacia auriculiformis*, litter cover is more important in reducing soil erosion than is the cover provided by standing biomass.
13. For summaries of the meaning of agroecosystem buffer capacity as concept, its elements and its management, the reader is referred to Budelman, 1991, and Budelman and Huijsman, 1991, pp. 38-39.

References

Adejuwon JO, Adesina FA (1990) Organic matter and nutrient status of soils under cultivated fallows: An example of Gliricidia sepium fallows from South Eastern Nigeria. Agroforestry Systems 10(1): 23-32

Agboola AA, Wilson GF, Getahun A, Yamoah CF (1982) Gliricidia sepium: A possible means to sustained cropping in Western Nigeria. In: MacDonald LH, ed, Agroforestry in the African humid tropics, pp 141-143. Tokyo: United Nations University

Aken'ova ME, Atta-Krah AN (1986) Control of Speargrass in an alley-cropping fallow. NFTA Research Reports 4: 27-28

Allen ON, Allen EK (1981) The Leguminosae. A source book of characteristics, uses and nodulation. Madison, USA: University of Wisconsin Press

Anonymous (1970) Les derniers résultats obtenus par la recherche agronomique dans la culture du vanillier. Cahiers d'Agriculture Pratique des Pays Chauds 1970-2: 101-105

Anonymous (1988) Faidherbia albida (Del.) A. Chev. Monographie. Nogent sur Marne, France: CTFT/CIRAD

Ashley J (1980) The culture of vanilla in Uganda. World Crops 32(5): 124-129

Beer J (1987) Advantages, disadvantages and desirable characteristics of shade trees for coffee, cacao and tea. Agroforestry Systems 5(1): 3-13

Beer J (1988) Litter production and nutrient cycling in coffee (Coffea arabica) or cacao (Theobroma cacao) plantations with shade trees. Agroforestry Systems 7 (2): 103-114

Belsky AJ, Amundson RG, Duxbury JM, Riha SJ, Ali AR, Mwonga SM (1989) The effects of trees on their physical, chemical and biological environments in a semi-arid savanna in Kenya. Journal of Applied Ecology 26: 1005-1024

Bene JG, Beall HW, Côte A (1977) Trees, food and people. Ottawa: IDRC

Benge MD (1981) Leucaena 3: The tree that is called saviour. World Crops 33(6): 132-133

Boissezon P de (1970) Etude du complexe absorbant des sols ferralitiques forestier de Côte d'Ivoire. Cahier ORSTOM, Série Pédologie 8(4): 391-416

Brams E (1971) Continuous cultivation of West African soils: Organic matter diminution and effects of applied lime and phosphorus. Plant and Soil 35: 401-414

Brewbaker JL (1987) Leucaena: A multipurpose tree genus for tropical agroforestry. In: Steppler HW, Nair PKR, eds, Agroforestry, a decade of development, pp 289-323. Nairobi: ICRAF

Budelman A (1988a) The performance of the leaf mulches of Leucaena leucocephala, Flemingia macrophylla and Gliricidia sepium in weed control. Agroforestry Systems 6(2): 137-145

Budelman A (1988b) Leaf dry matter productivity of three selected perennial leguminous species in humid tropical Ivory Coast. Agroforestry Systems 7(1): 47-62

Budelman A (1988c) The decomposition of the leaf mulches of Leucaena leucocephala, Gliricidia sepium and Flemingia macrophylla under humid tropical conditions. Agroforestry Systems 7(1): 33-45

Budelman A (1989a) Nutrient composition of the leaf biomass of three selected woody leguminous species. Agroforestry Systems 8(1): 39-51

Budelman A (1989b) The performance of selected leaf mulches in temperature reduction and moisture conservation in the upper soil stratum. Agroforestry Systems 8(1): 53-66

Budelman A (1989c) Effect of the application of the leaf mulch of Gliricidia sepium on early development, leaf nutrient contents and tuber yields of water yam (Dioscorea alata). Agroforestry Systems 8(3): 243-256

Budelman A (1989d) Flemingia macrophylla (Willd.) Merr. In: Westphal E, Jansen PCM, eds, Plant resources of South-East Asia. A Selection, pp 128-129. Wageningen, The Netherlands: PUDOC

Budelman A (1990a) Woody legumes as live support systems in yam cultivation. 1. The tree-crop interface. Agroforestry Systems 10(1): 47-59

Budelman A (1990b) Woody legumes as live support systems in yam cultivation. 2. The yam–Gliricidia sepium association. Agroforestry Systems 10(1): 61-69

Budelman A (in press, 1991) Buffers beter benutten. Het belang van systeembuffers voor duurzaam landgebruik in de tropen. Landbouwkundig Tijdschrift 103(10)

Budelman A, Huijsman A (1991) Towards a systems perspective. In: Savenije H, Huijsman A, eds, Making haste slowly: Strengthening local environmental management in agricultural development, pp 35-47. Amsterdam: Royal Tropical Institute

Budelman A, Pinners ECM (1987) The value of Cassia siamea and Gliricidia sepium as 'in situ' support systems in yam cultivation. Experiences from a farm-based trial. Proceedings of the International Workshop 'Gliricidia sepium (Jacq.) Walp.: Management and Improvement', pp 90-91. Turrialba, Costa Rica: CATIE/NFTA

Budelman A, Zander PM (1990) Land-use by immigrant Baoulé farmers in the Tai region. South-west Côte d'Ivoire (Ivory Coast). Agroforestry Systems 11 (2): 101-123

Budowski G, Kass DCL, Russo RO (1984) Leguminous trees for shade. Pesquisa agropecuaria brasileira. Brasilia, 19 s/n: 205-222

Carlowitz PG von (1986) Multipurpose tree yield data – their relevance to agroforestry research and development and the current state of knowledge. Agroforestry Systems 4(4): 291-314

Charreau C, Vidal P (1965) Influence de l'Acacia albida Del. sur le sol, nutrition minérale et rendements des mils Pennisetum au Sénégal. Agronomie Tropicale 6-7: 626-660

Clarke WC (1976) Maintenance of agriculture and human habitats within the tropical forest ecosystem. Human Ecology 4(3): 247-259

Clayton WD (1958) Secondary vegetation and the transition to savanna near Ibadan, Nigeria. Journal of Ecology 46(2): 217-238

Conklin HC (1957) Hanunoo agriculture. A report on an integral system of shifting cultivation in the Philippines. Rome: FAO

Coster Ch (1932) Wortelstudien in de tropen. I. De jeugdontwikkeling van het wortelstelsel van een zeventigtal boomen en groenbemesters. Tectona 25 (2): 828-872

Dommergues YR (1987) The role of biological nitrogen fixation in agroforestry. In: Steppler HW, Nair PKR, eds, Agroforestry, a decade of development, pp 245-271. Nairobi: ICRAF

FAO (1985) Changes in shifting cultivation in Africa. Seven case studies. FAO Forestry Paper 50(1) (vide pp 27-54 and 81-108; 141-162). Rome: FAO

Felker P (1978) State of the art: Acacia albida as a complementary permanent intercrop with annual crops. USAID Grant nr. 211 (d). Riverside, USA: University of California, Department of Soil and Environmental Sciences

Francis PA, Atta-Krah AN (1989) Sociological and ecological factors in technology adoption: Fodder trees in South-east Nigeria. Experimental Agriculture 25(1): 1-10

Garrity DP (mimeograph, 1989) Hedgerow systems for sustainable food crop production on sloping lands: elements of a strategy. Paper presented at the Symposium on the development of upland hilly areas in region XII, Bansalan, Davao de Sur, December 20-22, 1989. Los Banos, Laguna, Philippines: IRRI

Grove AT (1951) Land-use and soil conservation in parts of Onitsha and Owerri Provinces. Geological Survey of Nigeria Bulletin 21. Zaria, Nigeria (vide pp 56-58)

Hairiah K, Noordwijk M van (mimeograph, 1986) Root studies on a tropical ultisol in relation to nitrogen management. Haren, The Netherlands: Instituut voor bodemvruchtbaarheid. Rapport 7-1986

Halos SC (1980) Ipil-ipil. The wonder ceases. Canopy International 6 (9): 7-9

Hladik A, Bahuchet S, Ducatillion C, Hladik CM (1984) Les plantes à tubercules de la forêt dense d'Afrique centrale. Revue d'Ecologie (Terre et Vie) 39: 249-290

Huxley PA (1985) The tree-crop interface or simplifying the biological environmental study of mixed cropping agroforestry systems. Agroforestry Systems 3(3): 251-266

Huxley, PA (mimeograph, 1986) Rationalising research on hedgerow intercropping. An overview. ICRAF Working Paper 40. Nairobi: ICRAF

Huxley PA, Mead R (mimeograph, 1988) An ecological approach to on farm experimentation. ICRAF Working Paper 52. Nairobi: ICRAF

Huxley PA, Darnhofer T, Pinney A, Akunda E, Gatama D (1989) The tree/crop interface: A project designed to generate experimental methodology. Agroforestry Abstracts 2: 127-145

IDRC (1982) Leucaena Research in the Asian-Pacific Region. Proceedings of a Workshop, Singapore, 23-26 November 1982. International Development Research Centre (vide p 173)

Jonsson K, Fidjeland L, Maghembe JA, Hoegberg P (1988) The vertical distribution of fine roots of five tree species and maize in Morogoro, Tanzania. Agroforestry Systems 6(1): 63-69

Kang BT, Wilson GF, Lawson TL (1984) La culture en couloirs. Un substitut d'avenir à la culture itinérante. Ibadan, Nigeria: IITA

Kellman M (1979) Soil enrichment by neotropical savanna trees. Journal of Ecology 67: 565-577

Kessler JJ, Breman H (1991) The potential of agroforestry to increase primary production in the Sahelian and Sudanian zones of West Africa. Agroforestry Systems 13(1): 41-62

Keuchenius AAMN (1927) Over de ontwikkeling van leguminosen wortels. Bergcultures 1: 461-466

King KFS (1979) Agroforestry. In: Agroforestry: Proceedings of the 50th Symposium on Tropical Agriculture. Bulletin 303. Amsterdam: Royal Tropical Institute

King KFS (1987) The history of agroforestry. In: Steppler HW, Nair PKR, eds, Agroforestry a decade of development, pp 3-11. Nairobi: ICRAF

Lemaître C (1954) Le Faidherbia albida. In: Felker P (1978) State of the art: Acacia albida as a complementary permanent intercrop with annual crops. USAID Grant nr. 211 (d), pp 46-50. Riverside, USA: University of California, Department of Soil and Environmental Sciences

Lévi-Strauss C (1950) The use of wild plants in tropical South America. In: Steward, JH (1950) Handbook of South American indians. Vol 6, pp 465-486. Smithsonian Institute, Bureau of American Ethnology Bulletin 143, Washington: US Government Printing Office

Manner HI (1981) Ecological succession in new and old swiddens of montane Papua New Guinea. Human Ecology 9(3): 359-377

Marinet J (1955) Etude économique et culturale du poivre au Cambodge. Agronomie Tropicale 10(3): 279-360

Maydell HJ von (1983) Arbres et arbustes du Sahel. Leurs caractéristiques et leurs utilisations. Eschborn, FRG: GTZ

Miracle MP (1967) Agriculture in the Congo basin. Tradition and change in African rural economies. Madison, Milwaukee, and London: The University of Wisconsin Press

Monteny B, Eldin M (mimeograph, 1977) Données agroclimatiques receuillis à la station ORSTOM Adiopodoumé: 1948-1976, Abidjan, RCI: ORSTOM

Nair PKR (1987) Agroforestry systems inventory. Agroforestry Systems 5(3): 301-317

NFTA (1988) Leucaena Psyllids – A review of the problem and its solutions. Waimanalo, Hawaii, USA: Nitrogen Fixing Tree Association

Oldeman RAA (1986) Biological implications of leguminous tree architecture. Invited Paper. Abstracts 2nd International Legume Conference, pp 3-4. St. Louis, Missouri, USA

Pelissier P (1966) Les paysans du Sénégal. Les civilisations agraires du Cayor à la Casamance (vide pp 265-274). St. Yrieix: Fabrègue

Raintree JB, Warner K (1986) Agroforestry pathways for the intensification of shifting cultivation. Agroforestry Systems 4(1): 39-54

Roose E (1986) Terrasses de diversion ou microbarrages perméables? Analyse de leur efficacité en milieu paysan ouest-africain pour la conservation de l'eau et des sols dans la zone soudano-sahélienne. Cahier ORSTOM, Série Pédologie 22 (2): 197-208

Roose E, Chéroux M (1966) Les sols du bassin sédimentaire de Côte d'Ivoire. Cahier ORSTOM, Série Pédologie 4(2): 51-92

Rouw A de (1987) Tree management as part of two farming systems in the wet forest zone (Ivory Coast). Acta Oecologica. Oecologia Applicata 8: 39-51

Rouw A de (1991) Rice, weeds and shifting cultivation in a tropical rain forest. A study of vegetation dynamics. PhD thesis, Agricultural University Wageningen, The Netherlands. Paris, France: privately published

Ruyssen B (1957) Le Karité au Soudan. L'Agronomie Tropicale 12: 142-172, 279-306, 415-440 (vide p 287)

Ruthenberg H (1976) Farming systems in the tropics. 2nd edition. Oxford: Clarendon Press

Sanchez PA (1976) Properties and Management of soils in the tropics. Chichester, UK: John Wiley and Sons

Sanchez PA (1977) Soil fertility and conservation considerations for agroforestry systems in the humid tropics of Latin America. In: Mongi HO, Huxley PA, eds, Soils Research in Agroforestry, pp 79-124 (vide p 99)

Sanchez PA (1987) Soil productivity and sustainability in agroforestry systems. In: Steppler HW, Nair PKR, eds. Agroforestry, a decade of development, pp 205-223. Nairobi: ICRAF

Sanchez PA, Salinas JG (1981) Low input technology for managing Oxisols and Ultisols in Tropical America. Advances in Agronomy 34: 279-406

Seibert B (1987) Pepper on living poles of Gliricidia sepium and Erythrina lithospermum in East Kalimantan. Proceedings of the International Workshop 'Gliricidia sepium (Jacq.) Walp.: Management and Improvement', pp 111-118. Turrialba, Costa Rica: CATIE/NFTA

Seignobos C (1982) Matières grasses, parcs et civilisations agraires (Tchad et Nord Cameroun). Cahiers d'Outre Mer 35 (139): 229-269

Seubert CE, Sanchez PA, Valverde C (1977) Effects of land clearing methods on soil properties and crop performance in a Ultisol of the Amazon jungle of Peru. Tropical Agriculture (Trinidad) 54: 307-321

Siaw DEKA, Kang BT, Okali DUU (1991) Alley cropping with Leucaena leucocephala (Lam.) De Wit and Acioa barteri (Hook. f.) Engl. Agroforestry Systems 14(3): 219-231

Siemonsma JS (1982) La culture du Gombo (Abelmoschus spp.). Légume-fruit tropical. Ph.D thesis, Agricultural University Wageningen, The Netherlands

Steppler HW (1987) ICRAF and a decade of agroforestry development. In: Steppler HW, Nair PKR, eds, Agroforestry, a decade of development, pp 12-21. Nairobi: ICRAF

Steppler HW, Nair PKR (1987) Agroforestry, a decade of development. Nairobi: ICRAF

Sumberg J, Okali C (1988) Farmers, on-farm research and the development of new technology. Experimental Agriculture 24(3): 333-342

Verdcourt B, Trump EC (1969) Common poisonous plants of East Africa. London, UK: Collins

Watters RF (1971) Shifting cultivation in Latin America. Rome: FAO

Wiersum KF (1985) Effects of various vegetation layers in an Acacia auriculiformis forest plantation on surface erosion in Java, Indonesia. In: El-Swaify SA, Moldenhauer WC, Lo A, eds, Soil erosion and conservation, pp 79-89. Ankeny, Iowa: Soil Conservation Society of North America

Wiersum KF, Anspach PCL, Boerboom JHA, de Rouw A, Veer CP (1985) Integrated Rural Development Project, Eastern Province, Sierra Leone. In: Changes in shifting cultivation in Africa, pp 27-54. FAO Forestry Paper 50(1). Rome: FAO

Wilken GC (1977) Integrating forest and small-scale farm systems in Middle America. Agro-Ecosystems 3: 291-302

Willey RW (1975) The use of shade in coffee, cocoa and tea. Horticultural Abstracts 45(12): 791-798

Willey RW (1979) Intercropping – Its importance and research needs. Part 1: Competition and yield advantages. Field Crop Abstracts 32(1): 1-10

Wilson GF, Akapa K (1981) Improving the 'in situ' stem support system for yams. In: Terry ER, Odura KA and Caveness F, eds, Tropical Root Crops Research Strategies for the 1980s. Proceedings 1st Triennial Root Crops Symposium of the International Society for Tropical Root Crops–Africa branch, 8-12 September 1980, pp 195-197. Ibadan, Nigeria

Wit CT de (1989) Problemen van duurzaamheid in de landbouw. Landbouwkundig tijdschrift 101(1): 18-20

Young A (1987) The potential of agroforestry as a practical means of sustaining soil fertility. ICRAF Reprint 38. Nairobi: ICRAF

II Leaf productivity and nutrient content of mulches

Agroforestry Systems 7: 47–62 (1988)

2
Leaf dry matter productivity of three selected perennial leguminous species in humid tropical Ivory Coast

ARNOUD BUDELMAN
Centre Néerlandais, Ivory Coast. Department of Tropical Crop Science, Agricultural University. P.O. Box 341 Wageningen, The Netherlands

Key words: Leucaena leucocephala, Gliricidia sepium, Flemingia macrophylla, Leaf dry matter productivity, NELDER-design, leaf area indices

Abstract. Leaf dry matter productivity has been measured at the end of regrowth cycles of 3 months for stands of *Leucaena leucocephala, Gliricidia sepium* and *Flemingia macrophylla*, all three species belonging to the Leguminosae. The effective period of observation was minimally 24 months, or 8 regrowth cycles, after a period of establishment of slightly more than a year. Using systematic trial designs, originally developed by NELDER, yield figures have been acquired for a broad range of plant densities. At a plant density of 10,000 trees per hectare the average yearly leaf DM productivity amounted to 15.4 ton for *Leucaena leucocephala*, 12.4 ton for *Flemingia macrophylla* and 10.5 ton for *Gliricidia sepium*.

Productivity figures and leaf area indices are compared with those of relevant cropping systems.

1. Introduction

This paper summarises leaf biomass productivity[1] data from *Leucaena leucocephala, Gliricidia sepium* and *Flemingia macrophylla*. These perennial species belong to the Leguminosae, and are known to potentially fix atmospheric nitrogen.

The three woody legumes have been studied in sub-equatorial Ivory Coast as candidate species for low external input agricultural systems, like for example alley-cropping [19], whereby the trees help to prevent degradation of land under cultivation. Such systems permit a more intensive use of the land and may eventually replace shifting cultivation if both their sustainability and economic viability can be unequivocally demonstrated.

The information presented here is part of a series of articles prepared for this Journal. Apart from the observations on biomass productivity the research programme included studies on i) the characteristics of the three

[1] In the text the term 'production' is reserved to indicate the biological process *sensu stricto*. When referring to the rate of production the term 'productivity' is used [20]

leaf mulches (leaf nutrient content [11] and decomposition [12]), ii) the effects of leaf mulches in the cropping environment (in weed control [10] and as green manure[2]) and, lastly iii) the relative suitability of the woody leguminous species as live support systems in yam cultivation [9][3].

Nitrogen-fixing trees and shrubs offer a wide variety of possible uses in tropical production systems involving crop-, animal- and (fire) wood production. Brewbaker et al. [1972] noted that *Leucaena leucocephala,* grown in pure, dense stands and well-managed, probably represents the most efficient way of producing proteins under tropical conditions [8].

The list of other uses is extensive; in landreclamation, as covercrops in tree plantations, in weed control, as planted fallow systems, as wind- and fire-breaks, for wood production, *etcetera.*

In fact the evaluation of leguminous species as a biological resource has only just begun. Of the three perennial species considered here only *Leucaena leucocephala* has so far been studied in detail [6, 8, 17, 19]. For *Gliricidia sepium* knowledge of its productivity is incomplete and fragmented at best. As recently as 1982 review articles on the species lacked a single reference to biomass productivity [16], or contained only rudimentary data [14].

Since then information on the productivity of *Gliricidia sepium* has been gathered in Central America, where the species is used as living fences in cattle-rearing systems [22]. This is, however, a rather exceptional situation, from which it is difficult to generalise about biomass yield data.

In the case of *Flemingia macrophylla* even less data are published. General references have been made to its use in dry-season browsing in natural stands [3], as planted forage species [4], as cover crop in experimental coffee plantations [7] and in a similar role in rubber cultivation [2].

The objective of this study was to acquire productivity data on the three perennials, as part of a broad evaluation of the usefulness of species for integration in sustainable food cropping systems. Information on leaf biomass productivity is necessary, for example, in order to assess species potential to recycle nutrients and to develop appropriate cropping system management procedures.

2. Location, soil and climate

The productivity trials were located in Ivory Coast near Abidjan (5°19′N, 4°13′W) on an experimental farm, 29 m above sea level. The climate in the

[2] In preparation: 'The effect of the application of the leaf mulch of *Gliricidia sepium* on early development, leaf nutrient content and tuber yields of Water Yam (*Dioscorea alata*)'
[3] Reference [9] is a preliminary report. A final paper is in preparation titled 'Woody legumes as live support systems in Yam cultivation' and is scheduled for submission mid-1988

coastal belt is sub-equatorial, with two rainfall peaks. The main rainy season occurs from May to July, with a lesser peak in October and November. June is the wettest month. The average annual precipitation over a 30 year period is 2131 mm, with a maximum of 3030 mm, and a minimum of 1621 mm [21].

The period from January to March is relatively hot and dry, with January as the driest month. It is then that the 'Harmattan', a dry wind originating in the desert to the north, may reach the coast. During this event, which differs from year to year in intensity and duration, radiation is reduced as a result of the dustload in the air. The relative air humidity may drop to a minimum of 65 percent. Maximum air humidity normally is 80%, with a peak of 90% during the rainy season. The daily potential evapotranspiration varies from 2.8 mm (June and July) to 4.7 mm (March). The periods December to March and August to September show water deficits (PET > rainfall).

The soil is formed on a small band of tertiary and quarternary deposits along the coastline. The dominant characteristic of these sandy, ferralitic soils is their strong desaturation combined with a weak base exchange capacity [23].

The total clay and silt content in the upper horizon typically amounts to 15% or less. Humus is mostly responsible for the adsorption complex. The nitrogen availability is directly related to the organic matter content in the soil (average C/N ratio equals 15). Generally the phosphorus content in the

Table 1. Summary of monthly climatic data at the trial site (Adiopodoumé, Ivory Coast. Modified after Monteny & Eldin, 1977)

Month	t min. °C	t max. °C	PET* mm month^{-1}	Hours sun shine	Global radiation 10^6 J m^{-2}	Average rainfall in mm	Air hum. min.%	Days with rain
Jan.	22.0	31.1	109	171	448	30	65	3
Feb.	22.9	31.9	114	178	501	66	66	4
Mar.	23.2	32.1	130	199	604	112	65	8
Apr.	23.2	31.8	123	188	573	143	67	10
May	23.0	31.0	115	175	556	295	70	16
Jun.	22.4	28.7	79	89	402	698	77	22
Jul.	21.7	27.7	83	96	354	268	76	13
Aug.	21.3	27.4	79	79	338	42	76	8
Sep.	21.8	28.0	82	87	393	77	73	12
Oct.	22.5	29.2	110	160	521	167	73	14
Nov.	22.5	30.5	116	189	495	152	70	13
Dec.	22.2	30.4	109	176	481	81	67	7
Total	–	–	1249	1787	5666	2131	–	130
aver.	22.4	30.0	–	–	–	–	70	–

* PET (potential evapo-transpiration) calculated according to Turc's formula

Table 2. Physical and chemical characteristics of the soil in the trial area (abbreviated after Siemonsma, 1982)

parts 50–2000 μm	84.5	percent
clay + loam (parts < 50 μm)	14.1	percent
organic matter	1.06	percent
carbon (Walkley & Black)	6.11	promille
nitrogen	0.59	promille
C/N ratio	10.3	
P total (HNO_3)	0.24	promille
P available (modified Olsen)	0.04	promille
calcium	0.29	Meq. 100 g soil
magnesium	0.12	Meq.
potassium	0.01	Meq.
sodium	0.02	Meq.
total;	0.44	Meq.
Cation Exchange Capacity	3.27	Meq.
Base Saturation	13.6	percent
pH H_2O	4.7	
pH KCl	4.2	

soil is adequate for crop growth. Occasionally, the soil may be deficient in magnesium, but it is potassium that is repeatedly found to be the nutrient limiting biomass production.

The pH is often variable at the surface and ranges from 3.9 to 5.7, with 5.0 as average.

In Table 2 the analysis results are presented from soil samples taken from the trial area.

Under a closed canopy the physical structure of these soils is excellent. Rapid degradation however occurs whenever they are subjected to annual cropping regimes, and severe erosion is not uncommon under such circumstances [23].

3. Materials and methods

Cuttings of *Gliricidia sepium* (Jacq.) Walp.[4] and seeds from *Flemingia macrophylla* (Willd.) Merr.[5] were acquired from the Coffee and Cacao Research Station at Divo, where both species are used in experimental coffee plantations as shade- and cover crops. The seeds of *Leucaena leucocephala* (Lam.) de Wit[6] were collected from naturalised specimens that occur widely in

[4] Synonym: *Gliricidia maculata*
[5] Synonyms: *Flemingia congesta, Moghania macrophylla*
[6] Synonym: *Leucaena glauca*

roadside vegetation in southern and central Ivory Coast. The plant type is comparable in *habitus* and behaviour to the shrubby 'Hawaiian' form, and was preferred as reference over *Leucaena leucocephala* cv. K-28, since the Hawaiian form yielded substantially more leaf biomass per unit time (48% more leaf DM per ha $year^{-1}$). This is contrary to the early findings in Hawaii where the K-series in general performed better than the local weedy form of the species [8].
Later work in Hawaii however gave results comparable to those found here [17].

The trials were established in late 1983–early 1984. After a period of establishment of about one year the young shrubs were completely stripped of all green material. This first cutting marked the beginning of the first production cycle.

Due to limited space available a systematic design (a so-called NELDER-design after its inventor) was chosen [5, 18]; see subscript Figure 1 for specifications. No replications were involved.

The plant densities thus varied from 1,200 to 14,800 plants per ha for *Flemingia macrophylla*, 1,200 to 48,000 plants for *Gliricidia sepium* and from

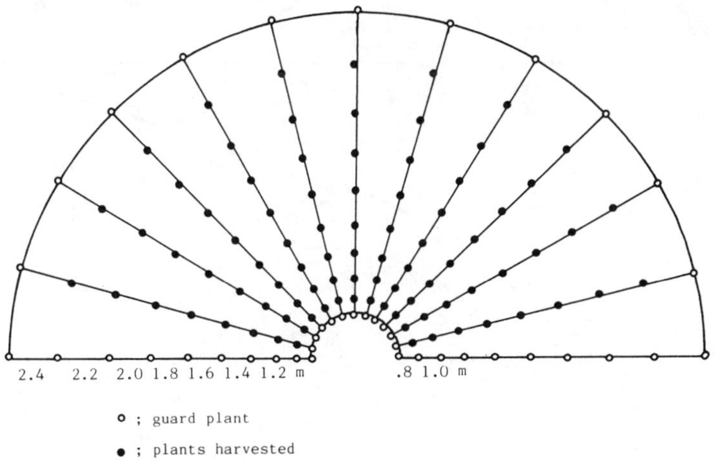

Fig. 1. Design of productivity trials.

Notes to Figure 1 – drawn to scale: example *Flemingia macrophylla* trial layout, 8 harvestable (half-) circles, 11 rays = 88 plants in observation.
- *Leucaena leucocephala;* as Figure 1 with 10 harvestable (half-) circles– 110 plants in observation.
- *Gliricidia sepium;* as Figure 1 with 10 harvestable (full-) circles – 220 plants in observation.

1,200 to 52,600 plants per ha for *Leucaena leucocephala*. The range for *Flemingia macrophylla* is less broad since the densest planted circles had to be abandoned during measurement, because the individual plants could no longer be identified.

In Figure 2 the deviations from square planting arrangements are given. Note that the deviations are slight in the upper ranges (> 5,000 plants ha^{-1}) and increase at the lower end of the plant density range. This deviation is important when comparing productivity data, since populations should not only be defined in terms of numbers per hectare, but also in terms of specific planting arrangements used [13, 27].

However, at lower densities the deviations from a square arrangement gradually become less important, since increasing distance means less competition between individual plants.

Leaf biomass productivity was measured after a standard 3-month regrowth period. The choice of the duration of the cycle was based on experiences elsewhere [6, 17].

At the end of the 90 day regrowth cycle the leaf biomass[7] per shrub was weighed, and a figure of yield per plant density was established by calculating the average weight of leaf biomass produced per shrub. Averages were

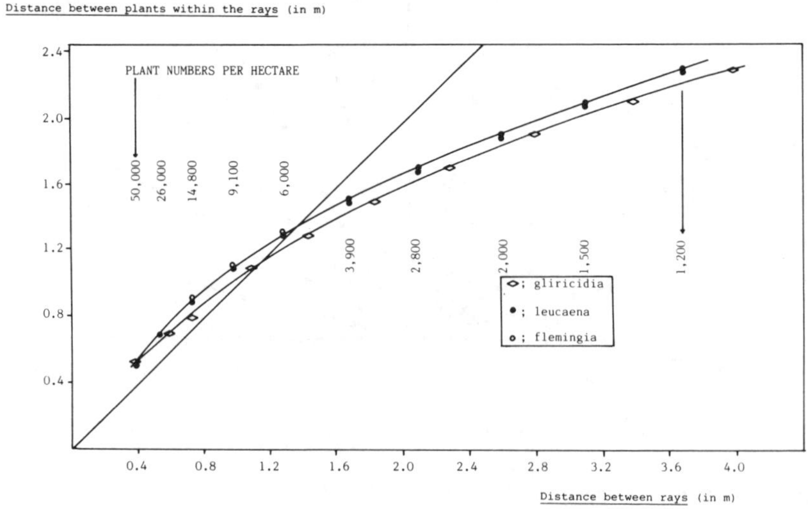

Fig. 2. Planting arrangements (n.b. the straight line represents square arrangements).

[7] A 'leaf' is defined as the complete compound leaf (including; *pulvinus, petiolus, rhachis, foliola*).

based on 22 plants (*G. sepium*), or 11 plants (*F. macrophylla, L. leucocephala*); Figure 1.

Averages were then converted into ton leaf dry matter per hectare. Dry matter data were calculated after drying samples at 105 °C

Regression analysis was subsequently applied to the data, using the natural logarithm of the plant densities on the X-axis. The resulting linear productivity functions are based on measurements of 220 (*G. sepium*), 110 (*L. leucocephala*) and 88 (*F. macrophylla*) shrubs harvested for each of the regrowth cycles; Figure 1. The 8 or 9 productivity functions acquired during the total period of observation were transformed into average productivity functions; Table 3.

The foliage, after cutting and fresh weight determination, was returned as mulch to the trial area. The data therefore come from a stable system from the point of view of plant nutrients available.

A standard cutting height of 50 cm was adopted, to permit fast regrowth [14]. In the course of time, for both *Gliricidia sepium* and *Leucaena leucocephala*, at all plant densities, trunk formation took place. The trees' architecture resembled that of a pollarded common willow (*Salix sp.*), with a few rounded thickening stumps from where the regrowth occured. *Flemingia macrophylla* is more tussocky in nature, and showed no rounded heads. In the latter case, many branches form basal clusters from where the sprouting starts.

During regrowth little weed control was needed, partly because of the speed of regrowth, partly because of the cut biomass acting as mulch preventing weed development [10].

The leaf surface to weight ratios were measured with a planimeter (model 'Milliplan' of METRAPLAN S.A.). These ratios were used to calculate the respective Leaf Area Index (LAI) and the specific mass of the leaf material.

Meteorological data came from the bioclimatology department of ORSTOM, Adiopodoumé. The observational field was situated at 500 m from the productivity trials.

4. Results

Average leaf biomass productivity functions for the three species studied are presented in Table 3. In the last column the range of plant densities per hectare is given over which the functions are valid. The functions are based on 9 three-monthly sequential[8] growth cycles in the cases of *Leucaena*

[8] The term 'sequential' is used here in the sense of similar events coming one after the other

Table 3. Average leaf-biomass productivity functions for the woody perennials

(Y in ton leaf DM ha^{-1} per 3 months, X is ^{e}log plants ha^{-1}) Species	Average productivity function	Number of cycles	Range plants ha^{-1}
G. sepium	Y = 0.59X–2.81	8	1,200–48,000
F. macrophylla	Y = 0.94X–5.55	9	1,200–14,800
L. leucocephala	Y = 1.06X–5.90	9	1,200–52,600

leucocephala and *Flemingia macrophylla*. The productivity function of *Gliricidia sepium* is an average of 8 cycles. The difference in the total length of the period of observation between the species is caused by the fact that *Gliricidia sepium* was slower to establish itself [11].

The equations presented in Table 3 allow us to estimate leaf biomass productivity ranging from very dense stands, useful when intensively managed forage production is the objective, to extensive 1,000-trees per hectare farming operations, the likely outcome should small farmers take up the idea of using trees in annual cropping systems [1].

In Figure 3 the correlation coefficients belonging to the individual 3-monthly productivity functions are displayed. As the Figure shows there is a slight but noticeable deterioration in the 'goodness of fit'. This is particularly true for the *Gliricidia sepium* trial where the values of r^2 decline

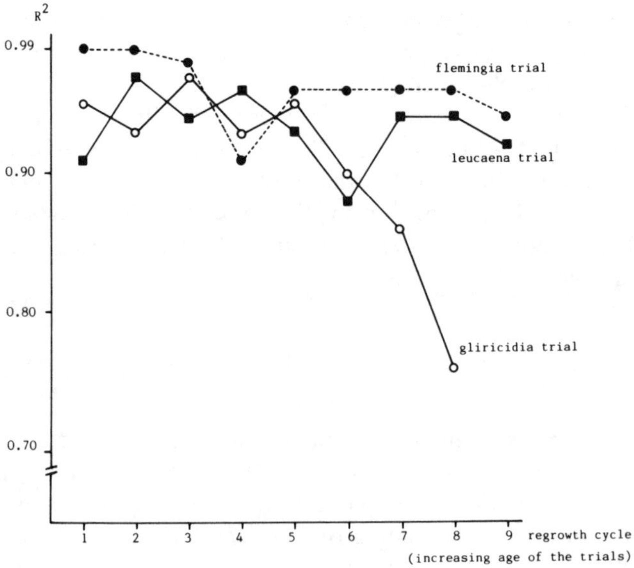

Fig. 3. 'Goodness of fit' (r^2) of the regression equations based on the leaf biomass harvested over the sequential regrowth cycles.

Table 4. Average leaf biomass productivity for a three-month regrowth period (at 10,000 plants ha^{-1})

Species	Quarterly leaf DM yield in ton and standard deviation	Value for *t*	Statistical significance
G. sepium	2.62 ± 0.63		
		1.58	N.S.
F. macrophylla	3.11 ± 0.53		
		3.01	$P < 0.010$
L. leucocephala	3.86 ± 0.60		
		4.18	$P < 0.001$
G. sepium	2.62 ± 0.63		

steadily. The reason is that in the course of time individual plants in the trial manage to enlarge at the expense of their neighbours. This is to be expected, and when distributed randomly in the trial no reason for immediate concern, since trial design and method of data processing smooth out such disparaties. The cause of deteriorating correlation coefficients is that whole circles, or large parts thereof, start to take over space belonging to adjacent circles.

The usefulness of the productivity trial ceases when the correlation coefficient falls below a pre-set value. Hence, for the *Gliricidia sepium* and *Leucaena leucocephala* trials, each involving 10 plant densities (Fig. 1), a r^2 value of 0.39 no longer meets the 95% confidence level minimally required for statistical significance [25; p. 184]. For *Flemingia macrophylla* with 8 plant densities a minimum r^2 value of 0.45 is required to meet the norm.

In Table 4 the average leaf biomass productivity for the three species has been calculated for a standard 10,000 plants per ha. The differences were tested using Student's *t*-test.

It follows from Table 4 that *Leucaena leucocephala* produces significantly more leaf dry matter per unit time than the other two species, whereas no statistically significant difference was found between *Gliricidia sepium* and *Flemingia macrophylla*. The yearly leaf DM yield for 10,000 plants ha^{-1} can thus be estimated at 15.4 ton for *Leucaena leucocephala,* 12.4 ton for *Flemingia macrophylla* and 10.5 ton for *Gliricidia sepium.*

5. Discussion and conclusions

Leaf dry matter productivity and rainfall

Quarterly leaf DM productivity varies considerably as the standard deviations in Table 4 show. The Figures 4-a to 4-c summarise evidence that the

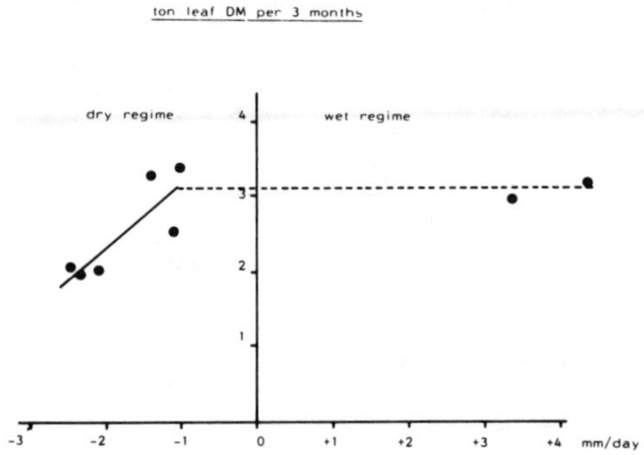

Fig. 4a. Leaf dry matter productivity of *Gliricidia sepium* as a function of the estimated average daily water availability (n.b. each dot represents the productivity of a 3-month regrowth period at 10,000 plants/ha).

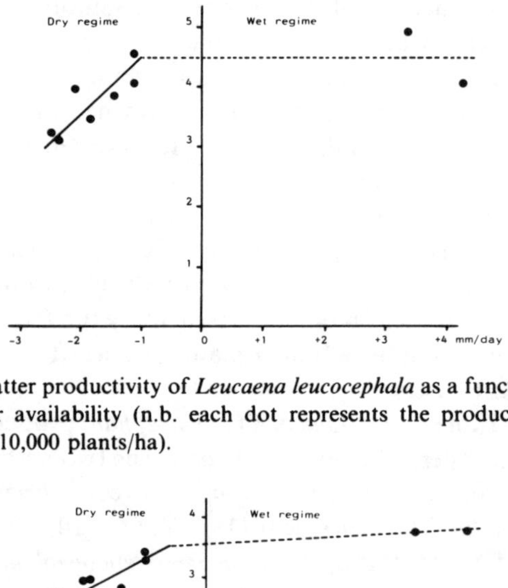

Fig. 4b. Leaf dry matter productivity of *Leucaena leucocephala* as a function of the estimated average daily water availability (n.b. each dot represents the productivity of a 3-month regrowth period at 10,000 plants/ha).

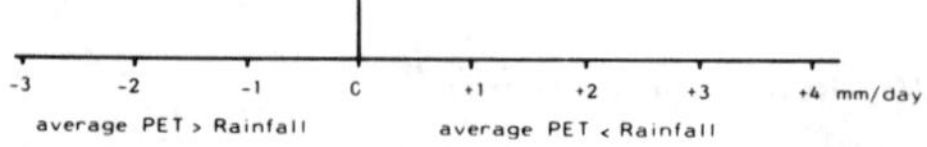

Fig. 4c. Leaf dry matter productivity of *Flemingia macrophylla* as a function of the estimated average daily water availability (n.b. each dot represents the productivity of a 3-month regrowth period at 10,000 plants/ha).

variation is caused by seasonal differences, with precipitation probably being the main factor.

Using the PENMAN formula for the Potential Evapotranspiration rate (PET) a crude average daily water-balance has been assessed for each of the regrowth periods. Climatic factors included in the calculations are: Radiation, windspeed, maximum- and minimum temperature, values for saturation vapour pressure and rainfall.

Two aspects potentially complicate the analysis. First of all the species studied are perennials and known to root deeply, with a possible access to permanent water sources.

Fluctuations in soil moisture content that affect tree biomass production as a function of rainfall are however likely to occur. The soils on the experimental site are freely draining and known to possess a limited water storage capacity only, because of their coarse, sandy texture and relatively low organic matter content; Table 2.

Moreover, two wells in the immediate neighbourhood had permanent water tables at a depth of over 60 m. The chance therefore that the plantations could have secured a permanent watersupply during the dry seasons is small.

A second fact that complicates the analysis is that crop water usage is not constant over the regrowth cycle. In the beginning, just after cutting, the water requirement will be much more limited compared to the situation at the end of the regrowth period when standing leaf biomass reaches a peak in terms of volume.

In the absence of any validated hydrological models for the species concerned, we opted for simple average daily water balance assessments. As Figures 4-a to 4-c show most production cycles experienced a water deficit (average daily potential evapotranspiration exceeding the average daily rainfall). During the 2 years of observation of productivity only during two production cyles the average daily precipitation exceeded the average daily value for the PET.

The productivity levels during 'dry regime' cycles, 6 for *G. sepium* and 7 for both *L. leucocephala* and *F. macrophylla* are clearly related to the average rainfall. The minimum confidence level for significance of the increase in productivity being 95% [25; p. 184]; Figure 4-a to 4-c.

Limitations of the approach

The main objective of this paper is to assess the performance of the three woody perennials in terms of leaf biomass productivity. For various reasons the degree of precision in the approach followed (and therefore the validity of the results acquired) is limited.

In spite of the fact that the productivity functions in Table 3 are specified for ranges up to a maximum of c. 50,000 plants ha^{-1} there are ample reasons to be careful with biomass productivity estimates in that specific part of the range. As Figure 2 shows the increase in plant numbers per ha in the upper part of the range is considerable, and in fact doubles between the two closest planted circles of shrubs in observation. As a 'safe' upper limit a maximum of 25,000 plants per hectare is advised for which the productivity estimates will have a realistic value.

Secondly, biomass was measured under a single cutting regime only, while plant densities and arrangements covered are a fraction of the planting configurations possible. Although the equations in Table 3 allow us to calculate productivity at dense, more or less equidistant planting arrangements (e.g. 20,000 plants ha^{-1}) the figure acquired is of doubtfull value when we intend to estimate the productivity of the same 20,000 plants arranged in an alley-cropping system in which for example the rows stand 4 m apart and the in-row plant distance is 12.5 cm.

When the productivity functions of Table 3 are used to estimate biomass productivity, careful consideration is needed regarding both aspects: Cutting regime and planting configuration.

Thirdly, the practical value of the equations from Table 3 is limited since productivity results are site-specific. Of the two main factors soil and rainfall, only regarding the latter factor, to some extent, crude corrections in productivity estimates can be made. With the information presented in Figure 4-a to 4-c, and knowledge on the local water-balance, the graphs allow modification of the average productivity figures given in Table 4.

Regarding soil differences such corrections to productivity estimates are impossible in the absence of verified models that relate nutrient uptake and biomass productivity of the species.

Comparison to other studies

Inversely, the same problem exists when reference is made to other studies and locations in the humid tropics.

As has already been argued, little information on biomass productivity is available for the three species, except for *Leucaena leucocephala*. Generally, data from literature on the performance of *L. leucocephala* concur with the results from this study [6, 8, 17, 19].

At the level of *interspecific* comparison of productivity data one meets the problem of different growthforms of the species considered (Table 5), a fact already relevant within the context of this study, since the shrubby *Flemingia macrophylla* definitely has a *habitus* that differs from that of the two tree

Table 5. The leaf dry matter productivity of the three selected species (at 10,000 plants ha^{-1}) compared to other mulch/fodder producing species
(periods of observation; 24 (*G. sepium*), 27 (*L. leucocephala* and *F. macrophylla*) and 20 months for the species studied by TALINEAU, 1970. All data come from the ORSTOM experimental farm, Adiopodoumé. Ivory Coast)

Species	Family	Number of harvests per year	kg DM ha^{-1} day^{-1}	Production relative to *Leucaena*
Panicum maximum	Gramineae	6.8	71.5 [26]	1.67
Cynodon sp.	Gramineae	7.0	51.8 [26]	1.21
Leucaena leucocephala	Leguminosae	4	42.8	1.00
Stylosanthes gracilis	Leguminosae	6.4	38.6 [26]	0.90
Flemingia macrophylla	Leguminosae	4	34.1	0.80
Gliricidia sepium	Leguminosae	4	29.1	0.68
Centrosema pubescens	Leguminosae	6.4	16.9 [26]	0.39

species. The shrub has less potential to expand compared to the tree species. As a consequence, at low plant densities, biomass productivity per unit area suffers; Table 7. A similar remark can be made concerning the comparison of the performance of the species studied with data on *crop associations* or, increasingly complex, *plant communities*; Table 6.

References at both levels are however of interest to acquire a more diverse idea of the performance of the leguminous perennials.

In Table 5 the leaf biomass productivity of the three leguminous species is compared to other mulch or fodder producing plant species, all grown in close stands. The data come from trials conducted on the same experimental farm at an earlier date. As Table 5 shows the two Gramineae have growth rates superior to the leguminous species. If the overall productivity is taken into account (that is including woody parts) the biomass produced by

Table 6. A comparison of the standing leaf biomass per unit area, the leaf area indices and the mean specific mass of the leaves of various plant communities
(data for the three species studied all at 10,000 plants ha^{-1}, 3 months old regrowth; additional data from Ewel et al. 1982; humid tropical Costa Rica)

Plant community	Standing leaf biomass $g\,m^{-2}$	LAI $m^2\,m^{-2}$	Leaf specific mass ($g\,m^{-2}$ of leaf)
sec. succession (11 months)	248	5.1	48 [15]
Gliricidia sepium	262	4.7	55
Flemingia macrophylla	311	6.7	47
cacao, plantain, *Cordia sp.*	312	3.4	91 [15]
Gmelina arborea	345	5.1	68 [15]
Leucaena leucocephala	386	7.5	52

leaf surface to weight ratio: leucaena 19.37 –; gliricidia 18.02 –; flemingia 21.45 $m^2\,kg^{-1}$.

Leucaena leucocephala for example would however probably equal that of *Cynodon sp.*

Increasing the plant density and the number of harvests per year may further increase the productivity of the three species studied. The Table provides evidence of the remarkable potential of *L. leucocephala* as a forage/mulch producing species.

Table 6 shows that *Leucaena leucocephala* pairs a high standing leaf biomass with a relatively low specific mass of the leaf material. The ratio gives an approximation of the consequently high Leaf Area Index.

Interestingly, because of its low leaf specific mass *Flemingia macrophylla* turns out to be an efficient soil cover.

Compare this situation to the cacao, plantain and *Cordia alliodora* association with approximately a similar standing leaf biomass per unit area, but with a much higher mean specific leaf mass. The Leaf Area Index found in the latter association is about half that of the stand of *Flemingia macrophylla.*

High Leaf Area Indices are obviously of importance when the species are planted to protect fields from the direct impact of raindrops and from soil erosion. The three leguminous perennials tested here however show the so-called nyctitropic movements, the repose of the leaflets in the absence of light. Analogous movements occur during heavy showers.

Of the three species studied *Leucaena leucocephala* shows the strongest movements, while *Flemingia macrophylla* is affected least of all.

This particular observation is important since the species' potential to provide proper soil protection will be considerably diminished during nighttime and/or showers. The LAI should therefore not be used uncritically as a single criterion in an evaluation of the actual soil cover provided by plant communities.

As a final conclusion Table 7 is presented, showing the yearly leaf biomass productivity estimates for the three species under different management scenarios, using the productivity functions of Table 3.

Table 7. Estimated yearly leaf dry matter productivity for the three species as a function of plant density (4 cuttings $yr.^{-1}$)

Management characteristics;	'extensive' ← (mixed cropping)			'intensive' → (monocropping)	
Species	1,000	5,000	10,000	15,000	25,000 pl ha^{-1}
Gliricidia sepium	5.1	8.9	10.5	11.4	12.6 ton
Flemingia macrophylla	3.8	9.8	12.4	13.9	–
Leucaena leucocephala	5.7	12.5	15.4	17.2	19.3

The extreme left column represents a smallholder operation in which the species are used in an extensive manner for feed production and/or poles. The trees under those circumstances will be planted at random and mixed with food- and other crops.

Under the extensive management scenario *Gliricidia sepium* is almost equally productive as *Leucaena leucocephala*, while *Flemingia macrophylla* clearly lags behind because of reasons explained previously.

Under an intensive management system, when planted as monoculture in dense stands, the superior production potential of *Leucaena leucocephala* is an obvious fact.

References

1. Agboola AA, Wilson GF, Getahun A and Yamoah CF (1982). Gliricidia sepium: A possible means to sustained cropping in Western Nigeria. In MacDonald LH (ed.) Agroforestry in the African humid tropics. United Nations University, Tokyo, Japan. p 141-143
2. Anon. (1964) Annual Report 1963-1964 of the Agricultural Research Station at Kade. Memorandum no. 3 University of Ghana.
3. Asare EO, Shehu Y and Agishi EA (1984) Preliminary studies on indigenous species for dry season grazing in the Northern Guinea Savanna zone of Nigeria. Tropical grasslands 18(3): 148-152
4. Bazill JAE (1987) Evaluation of tropical forage legumes under Pinus caribaea var. Hondurensis in Turrialba, Costa Rica. Agroforestry Systems 5(2): 97-108
5. Bleasdale JKA (1967) Systematic designs for spacing experiments. Experimental Agriculture 3: 73-85
6. Blom PS (1981) Leucaena, a promising versatile leguminous tree for the tropics. The International Tree Crops Journal 1: 221-236
7. Bouharmont P (1979) L'utilisation des plantes de couverture et du paillage dans la culture du caféier 'Arabica' au Cameroun. Café Cacao Thé 23(2): 75-100
8. Brewbaker JL, Plucknett DL and Gonzalez V (1972) Varietal variation and yield trials of Leucaena leucocephala (Koa Haole). Hawaii Agricultural Experimental Station, University of Hawaii Research Bulletin 166. pp. 29
9. Budelman A (1987) The above-ground structural compatibility of Flemingia macrophylla, Gliricidia sepium and Leucaena leucocephala as live stakes for Water Yam (Dioscorea alata). Proceedings of the International Workshop 'Gliricidia sepium (Jacq.) Walp.: Management and Improvement', 82-89. CATIE/NFTA June 21-27 1987, Turrialba, Costa Rica
10. Budelman A (1988) The performance of the leaf mulches of Leucaena leucocephala, Flemingia macrophylla and Gliricidia sepium in weed control. Agroforestry Systems 6: 137–145
11. Budelman A (in press) Nutrient composition of the leaf biomass of three selected woody leguminous species. Agroforestry Systems
12. Budelman A (in press) The decomposition of the leaf mulches of Leucaena leucocephala, Gliricidia sepium and Flemingia macrophylla under humid tropical conditions. Agroforestry Systems

13. Castillo AC, Moog FA and Avante DC (1979) Effects of row arrangements and plant density on herbage production and growth of Ipil-ipil. The Philippine Journal of Animal Industry 34: 75-81
14. Chadokar PA (1982) Gliricidia maculata. A promising fodder plant. World Animal Review 44: 36-43
15. Ewel J, Benedict F, Berish C, Brown B, Gliessman S, Amador M, Bermudez R, Martinez A, Miranda R and Price N (1982) Leaf area, light transmission, roots and leaf damage in nine tropical plant communities. Agro-Ecosystems 7: 305-326
16. Falvey JL (1982) Gliricidia maculata. A review. The International Tree Crops Journal 2: 1-14
17. Guevarra AB, Whitney AS and Thompson JR (1978) Influence of intra-row spacing and cutting regimes on the growth and yield of Leucaena. Agronomy Journal 70: 1033-1037
18. Huxley PA (1985) Systematic designs for field experimentation with multi-purpose trees. Agroforestry Systems 3: 197-207
19. Kang BT, Wilson GF and Lawson TL (1984) La culture en couloir. Un substitut d'avenir à la culture itinérante. IITA, Ibadan. Nigeria. pp. 22
20. Lieth H and Whittaker RH (1975) Primary productivity of the biosphere. Springer-Verlag, NY
21. Monteny B and Eldin M (1977) Données agroclimatiques receuillis à la station ORSTOM Adiopodoumé: 1948-1976. ORSTOM, Abidjan. RCI. pp. 24 (mimeogr.)
22. Otarola A, Martinez H and Ordonez R (1985) Manejo y produccion de cercas vivas de Gliricidia sepium en el noroeste de Honduras. Corporacion Hondurena de Desarollo Forestal/CATIE. Tegucicalpa, Honduras. pp. 24
23 Roose E and Cheroux M (1966) Les sols du bassin sédimentaire de Côte d'Ivoire. Cahier ORSTOM-Série Pédologie 4(2): 51-92
24. Siemonsma JS (1982) La culture du Gombo (Abelmoschus spp.) Légume-fruit tropical. Department of Tropical Crop Science, Agricultural University Wageningen, The Netherlands. pp. 297
25. Snedecor GW and Cochran WG (1973) Statistical methods (6th ed.). Iowa State University Press. pp. 593
26. Talineau JC (1970) Action des facteurs climatiques sur la production fourragère en Côte d'Ivoire. Cahier ORSTOM-Série Biologie 14: 51-76
27. Willey RW and Heath SB (1969) The quantitative relationships between plant population and crop yield. Advances in Agronomy 21: 281-321

Agroforestry Systems 8: 39–51, 1989.

3
Nutrient composition of the leaf biomass of three selected woody leguminous species

ARNOUD BUDELMAN
Centre Néerlandais, Abidjan. Côte d'Ivoire. Department of Tropical Crop Science, Agricultural University, Wageningen, The Netherlands
Present address: Royal Tropical Institute. Mauritskade 63, 1092 AD Amsterdam, The Netherlands

Key words: Leaf mulch, macro-nutrients, *Leucaena leucocephala, Gliricidia sepium, Flemingia macrophylla, F. congesta*

Abstract. In order to assess the value of leaf biomass the foliar nutrient content of *Leucaena leucocephala, Gliricidia sepium* and *Flemingia macrophylla* was studied under field conditions. In total 5 series of leaf samples were collected and analysed for nitrogen, phosphorus, potassium, calcium and magnesium content over a period of 42 months of growth observation. Results are compared with mean values derived from a literature study. Factors contributing to the variation observed are discussed and suggestions towards variation control are made.

1. Introduction

Leaf biomass harvested from deliberately planted trees and shrubs may be considered as an agricultural product or resource. Its use is generally confined to the sphere of small-farm operations. Depending on the type of biomass its utility-range includes options such as supplementary feeding of cattle or poultry, mulching for soil protection or green manuring to improve soil fertility. Leaf biomass as an agricultural resource becomes more important when modern, industrially-produced inputs are absent or too expensive.

The value of leaf biomass depends on the intrinsic properties of the material collected (nutrient content, the amino-acid composition of its protein, digestibility, resistence to decomposition) and its use.

Unlike commercially manufactured products (fertilisers, feedstuffs) leaf biomass carries no label giving specifications regarding its chemical composition. This poses problems in those cases in which biomass plays a vital role in the functioning of small-scale, low external input technology; alley-cropping, dry season feeding strategies, *etcetera*. Such systems while having the advantage of being based largely on local and accessible resources, can be neither understood properly nor improved unless a considerable research investment is done to determine the chemical properties of the biomass.

This paper, the second 'data-base' article[1] of a series of three, summarises information on the content of the 5 macro-nutrients (N, P, K, Ca and Mg) in the leaf biomass of *Leucaena leucocephala, Gliricidia sepium* and *Flemingia macrophylla* and the value of the biomass as a source of plant nutrients for use in low external input cropping systems.

The three woody leguminous species are known as potential fixers of di-nitrogen and were tested as candidates in live support systems for yam (*Dioscorea alata*) cultivation. Such tree-annual crop associations, in which the trees help to prevent degradation of land under cultivation, may contribute to ecologically sound sedentary land-use systems as alternatives to shifting cultivation.

2. Leaf nutrient content: a literature survey

The literature contains a range of information on the nutrient content of the leaves of *Leucaena leucocephala,* and to a lesser extent, of *Gliricidia sepium.* For *Flemingia macrophylla* only one reference was found; Table 1.

In general information on leaf nutrient composition in tropical leguminous trees and shrubs is scarce and incomplete. Sources are often inaccessible [1, 14, 15] and the majority of records does not give proper descriptions of the part(s) of the plant sampled, the age of the material, the analysis technique used, nor environmental information on ambient climatic and soil conditions.

Literature sources relate mainly to the use of foliage as feed stuff in animal husbandry research [4, 5, 6, 10, 23].

It will be clear from Table 1 that there is considerable variation in the figures presented by up to a factor 2.

3. Location, materials and methods

Leaf samples were collected from biomass productivity trials of *Leucaena leucocephala* (Lam.) de Wit, *Gliricidia sepium* (Jacq.) Walp.[2] and *Flemingia macrophylla* (Willd.) *Merr.*[3] located near Abidjan, Ivory Coast. After cutting and weighing the biomass for productivity figures, the leaf biomass was

[1] The first paper deals with aspects of leaf biomass productivity of the three species mentioned [2]. The third article investigates the decomposition of the leaf biomass, the relative persistency of the mulches and the release of nutrients [3]

[2] Synonym *Gliricidia maculata*

[3] Synonyms *Flemingia congesta, Moghania macrophylla*

Table 1. Nutrient content of the leaves of *Leucaena leucocephala, Gliricidia sepium* and *Flemingia macrophylla*; a literature survey

Nutrient	N	P	K	Ca	Mg	Ref.
species;						
Leucaena	2.87	0.10	—	2.22	—	[19]
leucocephala	3.12	0.16	2.00	2.00	0.40	[5]
	3.2*	0.27	2.91	0.84	0.29	[12]
	3.8–4.4	0.15–0.21	1.5–2.1	0.50–0.66	0.17–0.31	[10]
	3.84	—	—	—	—	[23]
	3.91	0.23	1.69	1.31	0.44	[13]
	4.2	—	—	—	—	[7]
	4.29	0.19	1.37	—	—	[15]
	4.31*	0.25	2.39	1.50	0.32	[8]
	4.33*	0.28	2.50	1.49	0.36	[14]
average;	*3.84*	*0.20*	*2.06*	*1.31*	*0.33*	
Gliricidia	3.01	0.11	—	0.60	—	[6]
sepium	3.46	—	—	—	—	[9]
	3.63	0.17	2.35	2.44	0.58	[5]
	3.7	—	—	—	—	[7]
	4.00*	0.28	3.33	1.60	0.42	[8]
	4.18	0.28	3.31	1.05	0.44	[4]
	4.21*	0.29	3.43	1.40	0.40	[14]
	4.39	—	—	—	—	[23]
	—	—	1.57*	—	—	[21]
average	*3.82*	*0.23*	*2.80*	*1.42*	*0.46*	
Flemingia macrophylla	3.79	0.25	1.04	0.65	0.28	[24]

* Young sprouts or prunings

returned to the soil as a mulch. The productivity trials were designed to function within a nutrient-stable context.

For information on the origin of the species in observation, climatic conditions and soil type see reference 2. The productivity trials were raised from material (seeds, cuttings) that received no inoculation with *Rhizobium* bacteria. All species showed root nodules, however.

The samples were taken over a period of 42 months. The sampling schedule is presented in Table 2.

A leaf sample contained leaflets and *rhachis*. Although it is likely that leaflets and *rhachis* differ in nutrient content, for practical purposes and for the sake of comparison with previously published information, both components were mixed and subjected to analysis.

Sampling was done at the end of a standard regrowth period of three months and the figures represent the average nutrient content of the foliar

Table 2. Number and timing of leaf samples taken

Time of sampling* (month)	Species		
	Leucaena leucocephala	*Gliricidia sepium*	*Flemingia macrophylla*
at 12	8	9	6
24	6	6	6
36	6	6	6
39	6	6	6
42 months	6	6	5
Total nr. of samples	32	33	29

* Months after planting the productivity trials

material of a *complete* branch. Young and old leaves were homogenously mixed in each sample.

Root nodules were collected and analysed for nitrogen from plants that were raised in 60 l steel drums; Table 6. The soil used in the drums came from an established plantation of *Leucaena leucocephala* cv. K–28. The root nodules were harvested from 6 months old plants.

The leaf- and nodule samples were analysed by the Central Laboratory in Ivory Coast of ORSTOM, the French organisation for overseas research and development co-operation.

The dry matter figures are derived from samples dried at 105° C; Table 3. Nitrogen was mineralised following the method developed by KJELDAHL and measured colorimetrically. The phosphorus in the samples was first changed into a phosphomolybdenum complex and then measured with the help of a colorimeter. Potassium content was quantified using a flame-

Table 3. Conversion factors; leaf dry matter (DM) weight as % of leaf fresh matter (FM) weight for each of the species

Species	Timing of sampling (month)					Mean and standard deviation
	12	24	36	39	42	
L. leucocephala leaf DM/leaf FM;	25.5	26.5	26.3	26.5	27.5%	26.5 ± 0.71
G. sepium leaf DM/leaf FM;	14.6*	21.3	21.2	22.3	22.7%	21.9 ± 0.74
F. macrophylla leaf DM/leaf FM;	28.5	28.6	27.6	28.9	28.1%	28.3 ± 0.25

* For *Gliricidia sepium* data at 12 months are excluded from calculating mean and standard deviation.

photometer. Lastly, calcium and magnesium levels were assessed by atomic absorption.

4. Results

Dry matter content

Conversion factors for the leaf dry/fresh weight are given for each of the species studied; Table 3. In the last column mean and standard deviation are given. In the calculation the data for *Gliricidia sepium* at 12 months were omitted, since by that time the species had not reached stable production at the management regime chosen [2]. It apparently develops slower than *Leucaena leucocephala* and *Flemingia macrophylla.*

While dry matter contents are approximately similar for *Leucaena leucocephala* and *Flemingia macrophylla,* the values for *Gliricidia sepium* are slightly lower. Literature sources, however, give higher values for leaf dry matter contents for *Gliricidia sepium*; 25.1 % [23], 27.0 % [4] and 27.1 % [6].

For *Leucaena leucocephala* the dry matter content of the leaves are in the range of 23.1 % [16] to 31.2 % [23], while for *Flemingia macrophylla* the only other study available gives 24.1 % for the leaf dry matter content [24].

Nutrient content

In table 4 the nutrient content of the leaves of the species studied is shown. Where a range is given the lefthand figure represents the estimate at 12 months, while the right hand figure gives the estimate at 42 months. The

Table 4. Nutrient content of the leaves of *Leucaena leucocephala, Gliricidia sepium* and *Flemingia macrophylla*

Nutrient	Content in percentages;		
	Leucaena	*Gliricidia*	*Flemingia*
nitrogen	2.95 to 3.67^a	3.18 ± 0.33	2.35 to 2.83^b
phosphorus	0.22 ± 0.03	0.19 to 0.25^c	0.19 to 0.25^d
potassium	1.52 ± 0.30	1.68 ± 0.20	0.98 to 1.40^e
calcium	0.92 ± 0.17	1.34 ± 0.18	0.65 ± 0.09
magnesium	0.31 to 0.37^f	0.40 to 0.61^g	0.20 ± 0.02

'upward trends' (average)

^{a}Y = +0.024 X + 2.66; P < 0.001: 3.37%
^{b}Y = +0.016 X + 2.16: P < 0.005: 2.62
^{c}Y = +0.002 X + 0.16: P < 0.010: 0.24
^{f}Y = +0.002 X + 0.28; P < 0.025: 0.34

'downward trends' (average)

^{d}Y = −0.002 X + 0.28; P < 0.001: 0.22%
^{e}Y = −0.014 X + 1.56; P < 0.001: 1.14
^{g}Y = −0.007 X + 0.69; P < 0.005: 0.48

Table 5. Statistical significance for the differences in the average nutrient content for the three species

Nutrient	*Gliricidia v. Leucaena*	*Gliricidia v. Flemingia*	*Leucaena v. Flemingia*
nitrogen	NS*	P<0.001	P<0.001
phosphorus	P<0.050	P<0.050	NS
potassium	P<0.050	P<0.001	P<0.001
calcium	P<0.001	P<0.001	P<0.001
magnesium	P<0.001	P<0.001	P<0.001

* No statistically significant difference between means.

regression equations at the bottom of the Table can be used to estimate the nutrient content for any other period between the two extremes (Y in % and X in months).

The confidence levels for statistical significance of the regression coefficients are calculated according to the method described in ref. 22; p.184.

In those cases where no trend could be established in the development of the nutrient content, a mean and a standard deviation is given.

In Table 5 the differences in mean nutrient contents found for the species are tested for significance using *Student*'s t-test.

In Table 6 information on the nitrogen content of root nodule samples is given and compared to the content of the nutrient in adjacent fine roots (< 1 mm). In the cases of *Leucaena leucocephala* and *Gliricidia sepium* sufficient material could be collected. Nodulation was rare with *Flemingia macrophylla*.

5. Discussion

5.1 Variation in nutrient composition

The literature study (see Table 1) makes clear that the leaf biomass nutrient content in the three species varies considerably. Regional differences in the

Table 6. The nitrogen content in the root nodules and adjacent fine roots for the three species at the age of 6 months

Species	N % in root nodules	N % in fine roots
Leucaena leucocephala	7.79 ± 1.18 (8)*	2.62 ± 0.34 (8)
Gliricidia sepium	7.45 ± 0.59 (8)	2.25 ± 0.33 (8)
Flemingia macrophylla	1.70 – 4.15 (2)	1.60 ± 0.25 (6)

* Between brackets the number of samples analysed. Each sample represents material from a single plant.

soil nutrient resources are probably the major factor behind differences in the leaf nutrient content. Jordan (1985; p. 55) has assembled evidence that tropical rainforest ecosystems differ widely in soil nutrient characteristics. For example, total soil nitrogen may range from as low as 785 kg ha^{-1} for a poor Venezuelan Amazon heath-forest up to 20,000 kg ha^{-1} in a rainforest covered soil with volcanic deposits in lowland Costa Rica. The differences in the figures for total phosphorus are even greater (36- against 7,000 kg P ha^{-1}).

In general the literature reflects higher average figures for the nitrogen and potassium content than does the present study. This holds as well for calcium in the cases of *Leucaena leucocephala* and *Gliricidia sepium*; Tables 1 and 4.

The lower mean values found for potassium and calcium in this study are probably the result of low availability of the elements. Potassium is known to be the most limiting element in the soils where the trials were conducted [18].

In the specific case of nitrogen the relatively low foliar content is more difficult to explain. In general the element is not considered as being a primary limiting nutrient in the soils of the trial area [18]. Moreover, at least in the cases of *Gliricidia sepium* and *Leucaena leucocephala* there is evidence that bacteria are active in di-nitrogen fixation in the root systems. The root nodules in the two species show nitrogen contents over 7 percent with limited variation: Table 6.

Root nodulation in *Flemingia macrophylla* is rarer than in the other two species and it is probable that the quantity of di-nitrogen actually fixed is low in *Flemingia macrophylla*. In older stands root nodules were much easier to be found in *Gliricidia sepium* and *Leucaena leucocephala* than was the case with *Flemingia macrophylla*. Table 4 shows however that the foliar nitrogen content in the latter species increases over time.

Apart from regional differences in soil nutrient resources and specific di-nitrogen fixation patterns other factors potentially contribute to the variation observed; i) the management system adopted (cutting frequency, use of the biomass produced), ii) overall age of the plants, iii) sampling procedure, and possibly iv) climatic influences.

Lower nutrient levels in this study are unlikely to result from depletion of soil resources as a function of the management system, since the leaf biomass was returned each time the biomass was cut [2]. It is probable that continual removal of the biomass produced would eventually lead to depletion of soil nutrient resources, resulting in lower levels of nutrients being available to plants.

The data presented in Table 4 do not provide reasons to assume that senescence of the plantations plays a role in decreasing nutrient contents. This was in any case anticipated in *Gliricidia sepium* and *Leucaena leucocephala*, since both are trees that can easily reach 20 years of age. During the period of observation (in total c. 45 months) the shrub *Flemingia macrophylla* did not show signs of senescence.

Nutrient content figures are likely to vary according to the part of the plant actually sampled. Complete shoots or prunings show lower nutrient levels compared to samples of foliar material only. Moreover, nutrient content -even on the same branch- varies significantly with the age of the leaves [19].

Seasonal changes contribute to the variation in foliar nutrient content. In the temperate zone particularly, temperature and precipitation are known to influence plant nutrient levels of trees [17]. Equatorial climates show limited temperature fluctuations, but rainfall varies significantly according to the seasons. However, within the limited period of observation of this study no correlations could be established between rainfall patterns and foliar nutrient content.

In almost all cases there exist significant inter-specific differences in the nutrient content; Table 5. *Gliricidia sepium* generally has the highest levels of leaf nutrient content, followed by *Leucaena leucocephala*, while *Flemingia macrophylla* systematically has the lowest levels. The latter species is apparently less demanding and may potentially perform better under situations of lower soil nutrient status than either *Gliricidia sepium* or *Leucaena leucocephala.*

Although *Gliricidia sepium* has higher levels of nutrients than *Leucaena leucocephala* (apart from nitrogen) the higher leaf dry matter productivity in the latter species ensures a higher annual yield of nutrients; Table 7. *Leucaena leucocephala can therefore be considered as a better means to recycle nutrients.*

Table 4 reflects the changes in the level of some of the nutrients during the period of observation. Statistically speaking, these changes in all cases are highly significant. There is, however, no systematic pattern in the direction of these changes, neither within species nor between species.

Increases, as they are found in the nitrogen content (*Leucaena leucocephala, Flemingia macrophylla*), the phosphorus content (*Gliricidia sepium*) and the magnesium content *(Leucaena leucocephala*) could be explained as a result of an accumulation process: The nutrients are collected in the deeper soil strata and arrive (when the biomass is cut and put as mulch around the stem) in the upper soil layer where they are subsequently trapped by the living root system and used for regrowth. This is potentially an important aspect in the dynamics of alley-cropping and agroforestry systems in general.

Table 7. Estimated annual nutrient yields in the leaf biomass of the three species (at 10,000 plants ha^{-1}, 4 cuttings yr^{-1}, amounts in kg ha^{-1})

Species kg	DM	N	P	K	Ca	Mg
Leucaena leucocephala	15,440	455–567	34	235	142	48–57
Gliricidia sepium	10,480	333	20–26	176	140	42–64
Flemingia macrophylla	12,440	292–352	24–31	122–174	81	25

In the specific case of an increase in the foliar nitrogen level increasing numbers of root nodules active in di-nitrogen fixation could be responsible.

Increases in nutrient levels are however neatly balanced by decreased levels; phosphorus and potassium (*Flemingia macrophylla*), and magnesium (*Gliricidia sepium*). The latter process is much harder to explain, partly because of the scattered and unsystematic nature of the phenomenon. Only in the case of potassium in *Flemingia macrophylla* does there exist a significant ($P < 0.025$) negative relationship between leaf biomass productivity and the foliar potassium content. Apparently the species can increase its leaf biomass volume without potassium becoming a limiting factor, whereas the nutrient is known as scarce in the soils of the trial area [18]. This observation illustrates the general conclusion made earlier that *Flemingia macrophylla* is less demanding in terms of soil fertility.

5.2. Leaf biomass as an agricultural resource

In Table 7 the quantities of nutrients mobilised annually are estimated by using average productivity figures collected during 9 (*Flemingia macrophylla, Leucaena leucocephala*) and 8 (*Gliricidia sepium*) regrowth cycles of three months each [2].

As Table 7 shows the leaf biomass yields of the three species do contain significant quantities of plant nutrients. As such these perennials function as accumulators rather than producers of nutrients, with the possible exception of some of the nitrogen.

Evidence brought together by Russell (1973) shows that the quantity of di-nitrogen actually fixed by bacteria is difficult to determine in the field and depends at least partly on the nitrogen content of the soil. Hence, on soils low in available nitrogen, bacteria in soybeans were found to fix substantially more nitrogen compared to fixation in the same crop on soils relatively rich in nitrogen [20; p. 381]. Since no information on the rate of di-nitrogen fixation is available for the plantations used in this study any estimate would be mere speculation.

However, the conclusion is in all cases that the better part of the nutrients in the leaf biomass originates from the nutrient pool in the soil, and that *the value of the perennials lies in making the nutrients available* for primary production purposes.

It is difficult to actually determine the economic value of the biomass produced by the woody legumes. One suggestion is to assess the *replacement value* of the leaf biomass produced, in terms of fertiliser or feed concentrate.

E.g., the nutrients accumulated in the annual leaf biomass yield of *Leucaena leucocephala, Gliricidia sepium* and *Flemingia macrophylla* (see Table 7) represent respectively 485, 365 and 305 US $ per ha at 1987 prices in Abidjan-port (Ivory Coast) for ureum, triple super-phosphate, potassium chloride (60%) and dolomite[4].

The comparison drawn has its flaws, however. Within the context of a farming system the application of fertilisers is an addition to the system and not simply a recycling of existing nutrients. Even if part of an added amount of fertilisers is leached out of the rooting zone normally explored by annual crops, the soil is richer than before the application and the nutrients form part of a reserve for future use when other, more deeply rooting crops can profit from its presence.

Secondly, the still unproven assumption here is that both sources of nutrients are equally efficient in terms of feeding a crop, and show similar rates of nutrient recovery. This is not necessarily true, and a comparative efficiency analysis (fertiliser *v.* mulch) is needed to determine the relative efficiency of mulches as source of plant nutrients. Such analysis is complicated by the fact that biomass releases its nutrients at a slower rate than ordinary fertilisers. Since nutrient release in decomposing biomass is dependent on rainfall, comparative trials must be repeated to include a rainfall range.

Evaluation of the feed value of the sources of leaf biomass (e.g. comparison with industrially processed feed concentrates) is equally complex. Palatability and digestibility differ according to species and relative contribution to feed mixtures, since, generally, leaf material from leguminous species is used as a feed supplement [5, 23]. In-vitro, *Gliricidia sepium* and *Leucaena leucocephala* are digested with equal ease [3, 5], but *Flemingia macrophylla* seems much less suitable as a supplementary source of feed [3; Table 3].

Whatever the use of the biomass may be, care should be given to the question *where* it is used. Table 7 shows that the three species have a notable potential to mine soil nutrient resources, while, like most of the oxisols in the humid tropics, the sandy sediments in littoral Ivory Coast contain little or no weatherable minerals, thus restricting nutrient reserves [18].

[4] 1 US $ = 303.3 Fcfa = 2.06 Dfl on September 23rd 1987

Table 8. Annual export of nutrients in the leaf biomass of the leguminous species relative to the nutrients removed in the tubers of a yam crop

Element[a]	Yam[b]	Leucaena[c]	Gliricidia[c]	Flemingia[c]
P	1.0	5.1	3.8	4.1
K	1.0	5.1	3.8	3.1
Ca	1.0	59.4	58.8	33.8
Mg	1.0	13.7	13.1	6.5

[a] nitrogen is not considered because in the case of the three leguminous species, part of it may be fixed, not having the soil as origin

[b] data yam tubers; 20,000 kg ha^{-1} fresh weight, DM 23.9%, P 0.14%, K 0.96%, Ca 0.05%, Mg 0.08% See: The effect of the application of the leaf mulch of *Gliricidia sepium* on early development, leaf nutrient content and tuber yield of Water Yam (*Dioscorea alata*). In press (this journal)

[c] based on data of Table 7

If biomass is systematically exported and used outside the land-use unit where it has been produced, tree leaf biomass productivity is likely to decrease and soil depletion imminent.

In Table 8 the nutrient quantity removed annualy when the leaf biomass of the leguminous species is used elsewhere, is related to the quantity that leaves the field in the harvested part of a food crop. Yam (*Dioscorea alata* cv. Brazo fuerte) is presented as reference crop, at a yield level of 20,000 kg ha^{-1} tubers, fresh weight. This food crop requires approximately 12 months from planting to harvesting.

Table 8 illustrates the fact that generally leaf biomass production requires considerably more nutrients per unit of dry matter compared to the production of carbohydrates. Therefore, under circumstances of limited soil nutrient resources, crop and/or animal production systems involving leguminous species should be *based on recycling rather than permitting the export* of nutrients. Alley-cropping maize with *Leucaena leucocephala* is a relevant example of such production system [12].

6. Conclusions

As has been shown, there is a considerable variation in the nutrient content of the leaf biomass of the three species studied; Table 1 and 4. This makes evaluation of leaf biomass as an agricultural product or resource difficult without costly foliar analysis. Still, an attempt to evaluate leaf biomass in monetary terms is necessary if, for example, alley-cropping system models are to include economic parameters.

The fundamental problem is that there are so many possible sources of variation, apart from the inter-specific differences in productivity and nutrient content given in the Tables 4 and 7.

Although variation in environmental conditions (e.g. soil nutrient content) will never be eliminated, the usefulness of data collection can be improved enormously when sampling procedures are standardised as far as possible, describing the exact parts and age of the material sampled. A summary on the climatic- and soil conditions, as well as the crop management system applied, would be useful collateral information.

References

1. Alferez AC (1980) Leucaena-based farming: Potentials and prospects. Animal Husbandry & Agricultural Journal (Philippines), July 1980; 18–29
2. Budelman A (1988) Leaf dry matter productivity of three selected perennial leguminous species in humid tropical Ivory Coast. Agroforestry Systems 7 (1): 47–62
3. Budelman A (1988) The decomposition of the leaf mulches of *Leucaena leucocephala, Gliricidia sepium* and *Flemingia macrophylla* under humid tropical conditions. Agroforestry Systems 7(1): 33–45
4. Carew BAR (1983) *Gliricidia sepium* as sole feed for small ruminants. Tropical grasslands 17(4): 181–184
5. Chadhokar PA (1982) *Gliricidia maculata.* A promising fodder plant. World Animal Review 44: 36–43
6. Gohl B (1975) Tropical feeds: feed formation summaries and nutritive values. FAO, Rome. pp. 188
7. IITA (1982) Annual Report 1981 (p. 29). International Institute for Tropical Agriculture, Ibadan. Nigeria. pp. 178
8. IITA (1982) Annual Report 1982 (p. 154). International Institute for Tropical Agriculture, Ibadan. Nigeria. pp. 217
9. Joachim AWR and Kandiah S (1934) The change in composition and decomposability of typical Ceylon green manures with age. Tropical Agriculturist (Ceylon) 82 (1): 3–20
10. Jones RJ (1979) The value of *Leucaena leucocephala* as a feed for ruminants in the tropics. World Animal Review 8: 13–23
11. Jordan CF (1985) Nutrient cycling in tropical forest ecosystems. Principles and their application in management and conservation. John Wiley & Sons, Chichester. UK. pp. 190
12. Kang BT, Wilson GF and Sipkens L (1981) Alley-cropping maize (*Zea mais* L.) and leucaena (*Leucaena leucocephala* Lam.) in Southern Nigeria. Plant and Soil 63: 165–179
13. Kleinjans JK (1984) Mineral composition of *Leucaena leucocephala* foliage. Leucaena Research Reports 5: 82
14. Koudoro D (1982) Evaluation of four woody fallow species for alley-cropping with maize and cowpeas. MSc thesis, National University of Benin, Cotonou. Benin (quoted in Kang BT, Wilson GF, Lawson TL, 1984. La culture en couloirs. Un substitut d'avenir à la culture itinérante. IITA, Ibadan. Nigeria. pp. 22)
15. Lao Lao MB (1978) (unpublished thesis UPLB, Philippines, quoted by Alferez AC, 1980. See ref. 1)
16. Lu CM and Hu TW (1981) Biomass production of two year old spacing trial plantation of Leucaena in Taiwan. Leucaena Research Reports 2: 53–55

17. Miller WF (1966). Annual changes in foliar nitrogen, phosphorus and potassium levels of Loblolly pine (*Pinus taeda* L.) with site and weather factors. Plant and Soil 24 (3): 369–378
18. Roose E and Chéroux M (1966) Les sols du bassin sédimentaire de Côte d'Ivoire. Cahier ORSTOM, Série Pédologie 4(2): 51–92
19. Rosas H, Quintero SO and Gomez J (1980) Nutrient evaluation of the arboreous legume Leucaena in Panama. Leucaena Newsletter 1: 18
20. Russell EW (1973) Soil conditions and plant growth. Longman, ELBS. 10th ed. pp. 849
21. Sivasangaranathan M and Thenabu MW (1978) Potassium content of *Gliricidia maculata* leaves and tender stems. Tropical Agriculturist 134: 109–111
22. Snedecor GW and WG Cochran (1973) Statistical methods. The Iowa State University Press. 6th ed. pp. 593
23. Vearasilp T (1981) Digestibility of rice straw supplemented with *Leucaena leucocephala* and *Gliricidia maculata*. Thai Journal of Agricultural Science 14: 259–264
24. Verlière G (1966) Valeur fertilisante de deux plantes utilisées dans les essais de paillage du Caféier: *Tithonia diversifolia* et *Flemingia congesta*. Café, Cacao, Thé 3: 228–236

III Decomposition and effects of the leaf mulches on the biophysical environment

Agroforestry Systems 7: 33–45 (1988)

4
The decomposition of the leaf mulches of *Leucaena leucocephala, Gliricidia sepium* and *Flemingia macrophylla* under humid tropical conditions

ARNOUD BUDELMAN
Centre Néerlandais, Abidjan, Ivory Coast. Department of Tropical Crop Science, Agricultural University. P.O. Box 341, Wageningen, The Netherlands

Key words: Leaf mulches, decomposition, *Leucaena leucocephala, Gliricidia sepium, Flemingia macrophylla,* nutrient release

Abstract. The decomposition of the leaf mulches of *Leucaena leucocephala, Gliricidia sepium* and *Flemingia macrophylla* has been studied under field conditions. Mulch samples were analysed over a period of 70 days on carbon, nitrogen, phosphorus, potassium, calcium and magnesium left. Loss rates for the elements show similar patterns for the three mulches. Losses of mulch material are caused by decomposition as well as removal by soil-dwelling fauna (insects, arthropods, etc.) feeding on the mulch.

Based on the data decomposition- and nutrient release functions are developed for the three mulches. These functions are useful in alley-cropping system analysis and models.

The decomposition characteristics of the three selected leaf mulches (data summary, rounded figures)

Mulchtype	Half-life value of the mulch	Persistency relative to *Leucaena* mulch	in vitro digestibility relative to *Leucaena*
Leucaena leucocephala	31 days	1.0	1.0
Gliricidia sepium	22	0.7	1.0
Flemingia macrophylla	53	1.7	0.5

1. Introduction

A mulch layer in a cropping system behaves in a similar fashion as litter layers found in natural ecosystems. The organic material gradually decays and partly becomes incorporated into the topsoil. In that process nutrients in the decomposing leaf mass are released and added to the nutrient stock in the upper strata of the soil. If the release is fast nutrients may become lost through leaching or volatilization (e.g. nitrogen in the form of ammonia). On the other hand if the rate of nutrient release is slow annual crops may not be able to profit from the nutrients available in a mulch layer applied at the beginning of the cropping cycle.

If no further additions take place the decomposition process in time results in less material, consequently diminishing the capacity of the layer to dampen soil temperature fluctuations, to prevent the germination of weed seeds, to limit soil erosion or to keep the soil from drying out.

All aspects have some importance at some time in cropping system functioning. When the behaviour of a certain mulch material is known, mulching like any other cultural practice, can be optimised so to perform beneficial functions at critical stages in the life-cycle of crops.

Whatever the reasons are behind mulch application it will be clear that knowledge on the rate of decomposition of the material is needed. The resistance of an organic material against decomposition, or its *persistency*, is a function of many factors. First of all there is an intrinsic factor, the specific properties of the biomass involved. Since decomposition can be seen as digestion of a food stock by a host of meso- and micro-organisms, the differences in decomposition rates in various materials are related to the quality of the mulch as food. High protein, high-nutrient and low-fiber material is obviously more attractive as a substrate and easier to digest than low-protein and fibrous material such as straw.

Apart from the material-specific aspects there are environmental factors at work that co-determine decomposition. Field studies [8,9,10] suggest that decomposition rates in the tropics are generally higher compared to rates found under temperate circumstances. The optimum situation for fast decomposition is where high average temperatures are found together with a continuous moisture supply. It follows that decomposition rates are highest in the continually humid lowland tropics [10].

This paper describes the results of a decomposition trial including the leaf mulches of *Leucaena leucocephala, Gliricidia sepium* and *Flemingia macrophylla.* The three woody perennials are studied in a role as live support systems in yam (*Dioscorea alata*) cultivation.

In the study a sub-set of questions relates to the application of the biomass produced by the shrubs as a mulch, useful in weed control, soil moisture conservation and soil fertility maintenance. An appreciation of the mulches in those respects requires knowledge of the decomposition characteristics of the materials.

2. Location and climate

The decomposition trial was conducted on the experimental farm of the ORSTOM research station at Adiopodoumé, near Abidjan, Ivory Coast, at

5° 19′ N., and 4° 13′ W. The site is located in the coastal belt, where we find a humid tropical lowland climate. The mean temperature is 26.2° Celsius. The average maximum temperature is 30.0, and the minimum temperature 22.4 °C.

There are two distinct rainy seasons. The main season falls in the months May to July, with a minor season in October to November. June is the wettest month when precipitation reaches its climax with an average of 698 mm. The yearly average rainfall over a period of observation of 30 years amounts to 2131 mm, with a maximum of 3030 mm, and a minimum of 1620 mm [12]. Maximum air humidity normally is 80%, with a peak of 90% during the rainy season. The daily evapotranspiration ranges from 2.5 mm (June to September) to 4.1 mm (February to March).

The decomposition data to be presented come from an experiment that ran during the first part of the long rain season. During the period of observation a total of 442 mm rainfall was recorded, or an average of 6.3 mm per day. With an average daily evapotranspiration of 2.8 mm the trial was conducted under a *wet regime*. Data were provided by the department of bioclimatology of the ORSTOM Research Station, Adiopodoumé.

For a review on soil aspects see reference 3.

3. Material and methods

Seeds from *Leucaena leucocephala* (Lam.) de Wit were collected from the 'Hawaiian' type locally found in roadside vegetations. Seeds from *Flemingia macrophylla* (Willd.) Merr. and cuttings from *Gliricidia sepium* (Jacq.) Walp. were acquired from the Coffee and Cacao Research Institute (IRCC) at Divo, Central South Ivory Coast, where both plant species are used as shade and cover crops. The mulch used in the decomposition experiment came from productivity trials involving the three species [3].

The data concerning the decomposition of the leaf[1] mulches come from an experiment in which 15 batches of fresh leaf mulch material per species were left to decompose. The trial was located in an open area, without shade. Before implementation of the experiment the soil was superficially hoed and raked. The trial area was kept free of weeds during the period of observation. The batches of fresh mulch, each containing the equivalent of 100 g dry matter, were spread out in wooden square frames, that measured 50 × 50 cm. The frames were 13 cm high, and slightly pressed into the soil. The construction prevented unwanted removal of mulch material by wind or water.

[1] A 'leaf' is defined as the complete compound leaf (including *pulvinus*, *petiolus*, *rhachis* and *foliola*)

Table 1. Analysis schedule of the mulch samples for each of the three species

Moment of sampling / Element	C	N	P	K	Ca	Mg
After 0 days no. of samples;	3	3	3	3	3	3
14 days	3	1	1	1	1	1
28 days	3	1	1	1	1	1
42 days	3	1	1	1	1	1
56 days	2[2]	1	1	1	1	1
70 days	3	3	3	3	3	3
Total no. of samples analysed;	17	10	10	10	10	10

[2] From each of the mulches one sample accidentally was lost.

The frames thus installed in the open field, allowed direct contact of the mulch with soil and air. No fabrics were used, so to give larger decomposers such as soil-dwelling insects and arthropods unhampered access.

Per sampling date per species the mulch from three frames was carefully collected, dried at 105 °C and analysed on the carbon content. For reasons of economy only the sample that came closest in weight to the average of the three of a series was further analysed for its nitrogen-, phosphorus-, potassium-, calcium-, and magnesium content.

The obvious disadvantage of the method followed is the considerable amount of time needed to collect the finer fraction of the mulch during sampling. This is particularly true for *Leucaena leucocephala* with its small leaflets. Moreover, contamination of the mulch material by soil particles takes place asking for time-consuming manipulation to separate the mulch material from soil particles.

The entire experiment took 70 days to complete. Sampling was done every fortnight. Including the initial analysis of the fresh material six series were analysed, distributed as follows in time; Table 1.

All analyses in Table 1 were done at the ORSTOM Central analysis laboratory, using the standard methods adopted there [1]. For a summary on methods applied see reference 4.

A computer programme with curve-fitting capability was used to process the nutrient content data. Invariably, of the possible curves the exponential function showed the lowest residual value, and consequently the highest r^2. As a measure of the 'goodness of fit' of the actual data and the curve fitted the r^2 values are presented in the 3rd column of Table 2. Knowing the value of r, Student's t-test can be used [16:p. 184] to establish the level of statistical significance of the fitting; Table 2, 4th column.

On request of the author in-vitro digestibility tests on the three leaf mulches were undertaken by the *Institute for Lifestock feeding and Nutrition Research* at Lelystad, The Netherlands.

4. Results

The equation that can successfully be fitted to data concerning the disappearance from a mulch layer of organic material or its components is based on the empirical exponential decomposition constant k [5,6,7,11,13];

$$Y(t) = Y(o) * e^{-kt} \qquad \text{eq. } <1>$$

in which $Y(o)$ is the original amount of material applied and $Y(t)$ the amount left after a period of time t. If time t is given in days k is expressed in day^{-1}. In the equation e is the base of the natural logarithm ($e = 2.718$).

Table 2. Loss patterns of carbon and the major nutrients from the mulches of *Gliricidia sepium*, *Leucaena leucocephala* and *Flemingia macrophylla* (mulch applied at a rate of 4,000 kg DM ha^{-1})

Nutrient or element	Half-life value (days)	Decomposition or release constant k (day^{-1})	Correlation coefficient r^2	Statistical significance of fitting
Gliricidia sepium;				
potassium	11.0	0.063	0.99	$P < 0.001$
magnesium	16.5	0.042	0.92	$P < 0.001$
phosphorus	19.8	0.035	0.92	$P < 0.001$
carbon	21.9	0.031	0.97	$P < 0.001$
nitrogen	21.7	0.032	0.95	$P < 0.001$
calcium	28.9	0.024	0.73	$P < 0.025$
Leucaena leucocephala;				
potassium	12.6	0.055	0.97	$P < 0.001$
magnesium	24.7	0.028	0.87	$P < 0.010$
phosphorus	26.2	0.026	0.94	$P < 0.001$
carbon	30.7	0.022	0.96	$P < 0.001$
nitrogen	38.5	0.018	0.90	$P < 0.001$
calcium	46.2	0.015	0.62	$P < 0.050$
Flemingia macrophylla;				
potassium	22.3	0.031	0.95	$P < 0.001$
magnesium	38.5	0.018	0.95	$P < 0.001$
phosphorus	34.7	0.020	0.95	$P < 0.001$
carbon	53.4	0.013	0.93	$P < 0.001$
nitrogen	53.4	0.013	0.89	$P < 0.001$
calcium	69.3	0.010	0.91	$P < 0.001$

Table 3. Dry matter contents[3] and in-vitro digestibility[4] of the leaf material of the selected mulch types

species	dry matter (%) and standard deviation	in-vitro digestibility of organic matter (%)
Gliricidia sepium	21.9 ± 0.74	64.2
Leucaena leucocephala	26.5 ± 0.71	64.8
Flemingia macrophylla	28.3 ± 0.25	31.0

[3] See reference 4.

[4] percentage of the organic matter digested after treatment with buffered rumen fluid followed by a treatment with pepsinhydrochloric acid

A shortcoming of equation <1> is that there are no terms that account for temperature and humidity as factors of influence on the rate of decomposition. Information on the life-span of mulch-layers probably is highly site-specific. For that reason a precise description of the climatological circumstances under which the trials took place is important to include in reporting (see 'Location and climate').

Fitting the data on the nutrient losses from the samples to eq. <1> gives excellent 'goodness of fit' and consequent statistical significance of the fitting; Table 2 (3rd and 4th columns).

Carbon makes up the bulk (slightly less than 50%) of the elements present in the dry leaf biomass, and its 'half-life' value can be applied as a general indicator of the persistence of a mulch material. The 'half-life' value, the time it takes to lose half of the material or half of the amount of a certain nutrient from a mulch layer, can be calculated by setting $Y(t)/Y(o) = 0.5$, using the relevant value for parameter k from Table 2 (2nd column).

As Table 2 shows the mulches are short-lived, except for the mulch of *Flemingia macrophylla*. The difference in persistence found in the three mulches is only partly explained by the higher dry matter content in *F. macrophylla* leaves; Table 3. More telling is the slow decomposition under circumstances of in-vitro digestion, suggesting the possible presence of a high percentage of lignin in the leaves of *Flemingia macrophylla*.

5. Discussion

JANSSEN [5, 6, 7], discussing the decomposition of organic matter in soils in general, shows that over longer periods of time (in the order of years) the speed – expressed as the 'relative decomposition rate' k – at which organic materials decompose decreases.

Here, for practical purposes (see eq. 2) and in view of the fact that mulch layers have functional life-spans measured in weeks rather than years, decomposition rates are assessed using the simple fitting procedure previously presented with a constant average value for *k*; eq. <1> and Table 2.

A more fundamental reason to refrain from using JANSSEN's model is that his formula is based on the loss of carbon in the form of CO_2 ('dissimilation') only, whereas it is likely that the observed loss of mulch material partly must be attributed to soil-dwelling fauna that feeds on the mulch. In that case decomposition may take place outside the mulch layer as the unit of observation, for example in the soil directly underneath the mulch layer. Application of JANSSEN's model to explain losses in mulch layers would then result into a possible over-estimation of the actual rate of dissimilation, distorting the derived parameter in his single-parameter model [5,6,7].

The argument gains weight when comparing the relative loss rates of the individual elements from the mulch material.

It is generally accepted that nitrogen and phosphorus are mineralised at a slower pace than carbon when the ratio between C/N and C/P in a given substrate exceeds a certain critical value. In that case a relative scarcity of N and P occurs, and micro-organisms tend to immobilise both elements for use in their own metabolism. Under those circumstances the process of decomposition of organic matter by micro-organisms competes with plant growth for the limited amount of nitrogen and phosphorus available.

In the case of the C/N ratio the accepted critical value is about 10. Values higher than 12 to 13 indicate a shortage of nitrogen [5: p. 174]. Table 4 shows that the C/N ratio belonging to the mulch types in all cases exceeds the critical value of 10. The high C/N ratio found for *Flemingia macrophylla* is probably a factor explaining the persistency observed.

However, when comparing the data on the losses of nitrogen in Table 2, at least in the cases of *Flemingia macrophylla* and *Gliricidia sepium* nitrogen is lost at the same speed as is found for carbon.

The critical upper value for the C/P ratio is less well defined, but immobilisation of phosphorus is known to start at C/P values > 100 [5:p. 187].

In all three cases however phosphorus is lost at a faster rate than carbon; Table 2.

As suggested earlier the likely explanation of this paradox is that losses occur via soil-dwelling fauna feeding on the mulch and transporting important parts thereof. Preferential feeding (the juicier cell-mass rather than the leaf-nerves) then explains the relatively rapid loss of nitrogen and phosphorus as compared to carbon, since both elements are relatively under-presented in leaf-nerf material as compared to carbon.

Table 4. The carbon/nitrogen and carbon/phosphorus ratio of the three mulches (based on un-decomposed dry matter)

species	C/N	C/P
Flemingia macrophylla	21.0	214.5
Leucaena leucocephala	12.1	202.1
Gliricidia sepium	12.1	179.2

For mulch layers it can therefore be concluded that the decomposition constant from eq. <1> catches at least two different processes; losses through true dissimilation and secondly losses through transport by animal life from the mulch layer: $k_{<eq.1>} = k_{diss.} + k_{trans.}$.

Hence, the decomposition of a mulch layer should not be uncritically considered as the very beginning of decomposition of organic matter in the soil in general using a single model to describe all stages.

Within the broader context of a mulched crop however the nitrogen and phosphorus removed from the mulch through transport can be considered as being released from the mulch into the environment. The role of the fauna in that respect is that of an intermediate agent that helps in making nutrients available for plant growth[5].

Table 2 shows that the patterns of cat-ion release are similar for the three mulch materials studied; potassium is lost rapidly, while calcium tends to stay behind in the decomposing leaf-mass. As in the case of nitrogen and phosphorus here as well fauna is at work helping to release the cat-ions into the environment. The *differences* in half-life values found in the group of elements analysed are however explained when looking at the role the individual elements play in the chemistry of plant tissues [15].

Potassium and magnesium are relatively rapidly lost. Potassium is present as a cat-ion freely moving in the cell fluid, taking part in the synthesis of amino-acids and proteins. When cell membranes desintegrate potassium is easily washed out from the organic material. It is well-known that potassium is also leached from living leaves [10].

Magnesium is as well present as a free cat-ion and released in a similar fashion as potassium. Part of the magnesium however is a constituent of complex molecular structures (for example chlorophyll and pectins) and therefore probably somewhat slower to move from the decomposing leaf material.

The slow loss rate of calcium is less easily understood. In general, in the plant the nutrient is not very mobile and tends to lag behind in senescent

[5] Obviously, part of the nitrogen and phosphorus is (temporarily) immobilised in the tissues of the fauna involved. Ultimately however, the nutrients will be mineralised via waste-products of foodchains (faeces, excretion) and decomposition of dead bodies

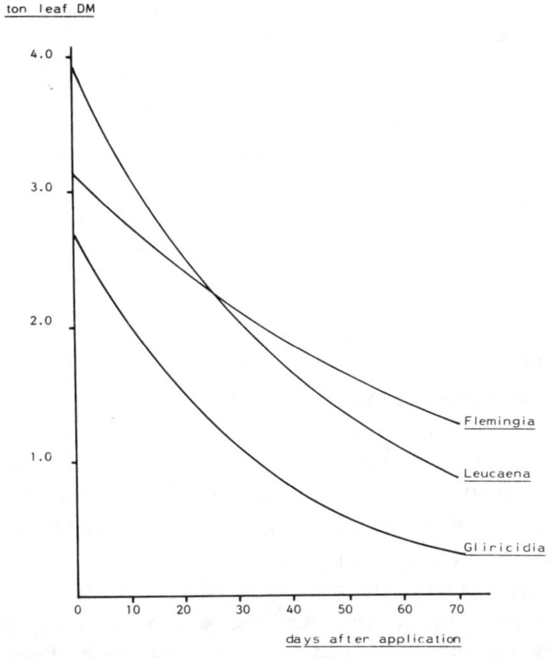

Fig. 1. **The decomposition of the mulches of *Leucaena leucocephala, Flemingia macrophylla* and *Gliricidia sepium* (note to Figure 1: Decomposition curves are based on information presented in Table 5).**

leaves. It is present in the form of Ca-pectate in the middle lamellae of the cellwall. Another possibility is that calcium is stored in the form of Ca-oxalate crystals. These concentrations may result in greater variations in the calcium content in the samples explaining the lower quality of the fitting of the calcium loss data as compared to the other elements; Table 2.

Application of the decomposition function

Knowledge on the properties of a mulch material, and especially its persistency is relevant in relation to its capacity to control weed development and to influence moisture and temperature gradients in the soil/air interface. Using the carbon loss rates or half-life values from Table 2 it can be concluded that, relative to the data for *Leucaena leucocephala*, the mulch of *Flemingia macrophylla* has a *'persistency factor'* of 1.75. For the mulch of *Gliricidia sepium* the factor is 0.71.

Table 5. Quarterly leaf biomass productivity of the three species (see Figure 1) and potassium content of the leaf mulches (see Figure 2)

Species	Average leaf biomass productivity[6] ton ha^{-1} per 3 months	Potassium content[7] in %	Estimated K yield kg/3 months
Gliricidia sepium	2.6 ton	1.68%	44 kg
Leucaena leucocephala	3.9	1.52	59
Flemingia macrophylla	3.1	1.19	37

[6] See ref. [3], at 10,000 plants ha^{-1}
[7] See ref. [4]

The data concur with observations from other decomposition studies. From evidence presented by READ et al (1985) one can conclude that the half-life value of fresh *Leucaena leucocephala* prunings is not much longer than three weeks. YAMOAH et al (1986) found that the mulch of *Gliricidia sepium* disappeared completely within 120 days. In the same trial, during the same period of observation, *Flemingia macrophylla* prunings lost 73% of the original amount of dry matter applied.

The loss of material from the mulches is illustrated in Figure 1. In order to put the usefulness of the mulches into a broader perspective decomposition is related to the *actual* leaf biomass productivity figures found for the three species [3].

The average leaf biomass productivity of *Gliricidia sepium* is relatively low, and combined with the lowest 'half-life' value (Table 2; figure for carbon) of the three mulches, its value in soil protection or weed control is limited; Figure 1.

Although from the point of view of productivity *Leucaena leucocephala* scores higher than *Flemingia macrophylla* (Table 5), the mulch of the latter species, given its greater persistence, will provide a more durable cover on the longer term; Figure 1. This difference in persistence partly explains the reasons behind the success of the mulch of *Flemingia macrophylla* in weed control [2].

Apart from the role of the mulches as layers that influence exchanges in physical processes (heat- and water transport, light) a mulch, in time, exerts an influence on the fertility of the soil. Plant nutrients gradually are freed during the decomposition process and become available for crop production. In that perspective a mulch functions analogously to a so-called *slow release fertilizer*.

The *decomposition equation* presented earlier can easily be inverted to obtain the *nutrient release function*:

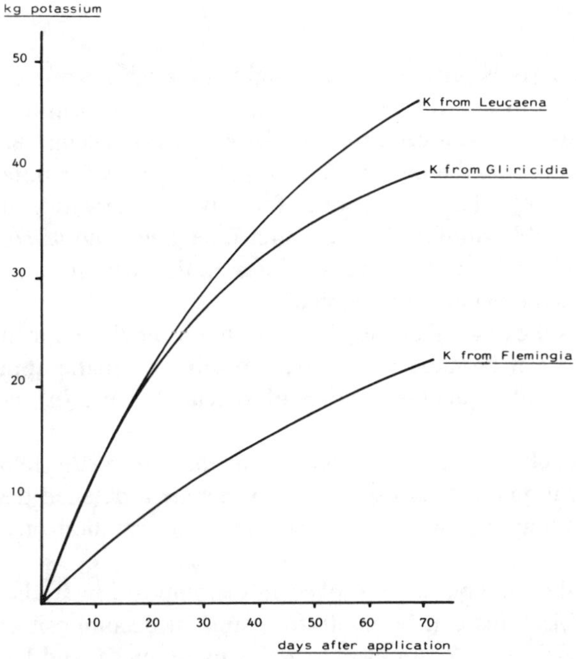

Fig. 2. The release of potassium from the mulches of *Leucaena leucocephala, Flemingia macrophylla* and *Gliricidia sepium* (note to Figure 2: release patterns based on information presented in Table 5)

$$R(t) = (D * P) * (1 - e^{-kt}) \qquad \text{eq. } \langle 2 \rangle$$

in which R(t) represents the amount of a specific nutrient released in kg ha^{-1} after a certain period of time (*t* in days), *D* is the mulch dry matter quantity initially applied in kg ha^{-1}, *P* is the presence of the nutrient in the mulch material as a fraction. The value $e = 2.718$, and for *k*, now the release constant, the relevant values of Table 2 can be adopted.

As an example of the use of eq. <2> we selected the element potassium. In Figure 2 the release patterns for potassium have been analysed. Here as well the graphs are based on *actual* information, in this case the estimated quantity of potassium in the quarterly leaf biomass yield for each of the three species; Table 5.

With information on the nutrient content of the leaf mulches available [4] the release patterns of the other major elements can be evaluated as well.

6. Conclusions

Decomposition trials provide useful insights in certain properties of woody species. The Figures 1 and 2 effectively demand a clarification of management priorities: "Do you need tree species efficient in accumulating potassium since you are searching for alley-cropping species for potassium-poor environments, or, is lasting soilcover for water conservation the issue because of unreliable rainfall?" In the first case *Leucaena leucocephala* and *Gliricidia sepium* will be obvious candidates, in the latter situation *Flemingia macrophylla* is more likely to succeed.

Whatever is the case, acquiring knowledge on the decomposition characteristics of the leaf biomass will be part of any systematic approach to an inventory and subsequent screening of species for use in alley-cropping systems.

Such approach will gain in efficiency if a standard methodology is adopted. Considerations in this respect are to include a detailed description of climatic conditions during trials, particularly information on rainfall and temperature.

Secondly, the methodology applied in decomposition studies should be evaluated. Objections can be made to results of decomposition studies in which prunings are sealed in bags made of nylon mesh, and left to decompose [17]. This method excludes a whole range of decomposing agents, such as larger insect species. Moreover, the use of bags shields mulch layers from the direct, destructive impact of raindrops during heavy showers. As the data in this study show soil-dwelling fauna potentially is a factor explaining loss patterns in mulch layers.

A third suggestion is to adopt a certain mulch type, for example that of *Leucaena leucocephala*, as a *standard* to interpret the results of comparative mulch decomposition studies. Such procedure is for example accepted in animal feed digestibility studies, and would be of help in comparing results from different sites and materials.

Eventually, alley-cropping studies will enter the field of modelling. It is simply inconceivable that the large bodies of data that now start to flow from inventories can be handled without the help of computer models. One future objective will be to simulate alley-cropping systems in order to predict and optimise their relative efficiency. In the complex exercise of grasping the dynamics of alley-cropping systems, decomposition- and nutrient release rates of mulches will be found as simple but essential parameters in models.

7. Acknowledgements

Thanks are due to Dr. B.H. JANSSEN and Mr. I.G.A.M. NOIJ for their help in interpreting the data. Mr. J.P.L. DE ROUW originally provided the computer software which greatly facilitated the curve-fitting routine.

8. References

1. Anon (undated) Brochure de Laboratoire Central d'Analyse. ORSTOM, Adiopodoumé. RCI. pp 16
2. Budelman A (1988) The performance of the leaf mulches of *Leucaena leucocephala, Flemingia macrophylla* and *Gliricidia sepium* in weed control. Agroforestry Systems 6: 137–145
3. Budelman A (in press) Leaf dry matter productivity of three selected woody perennial leguminous species in humid tropical Ivory Coast. Agroforestry Systems
4. Budelman A (in press) Nutrient composition of the leaf biomass of three selected woody leguminous species. Agroforestry Systems
5. Janssen BH (1983) Organische stof en bodemvruchtbaarheid. Caput Selectum in het kader van Bodemvruchtbaarheid en Bemesting. Vakgroep Bodemkunde en Plantevoeding. Landbouw Hogeschool, Wageningen, The Netherlands. pp 215
6. Janssen BH (1984) A simple method for calculating decomposition and accumulation of 'young' soil organic matter. Plant and Soil 76: 297–304
7. Janssen BH (1986) Een één-parametermodel voor de berekening van de decompositie van organisch materiaal. Vakblad voor Biologen 66 (20): 433–436
8. Jenkinson DS and Ayanaba A (1977) Decomposition of C-14 labeled plant material under tropical conditions. Soil Science Society of America Journal 41: 912–915
9. Jenny H, Gessel SP and Bingham FT (1949) Comparative study of decomposition rates of organic matter in temperate and tropical regions. Soil Science 68: 419–432
10. Jordan, CF (1985) Nutrient cycling in tropical forest ecosystems. Principles and their application in management and conservation. John Wiley & Sons, Chichester, UK. pp. 190
11. Monnier G and Gras R (1965) Action d'une culture d'engrais verts sur les propriétés physiques d'un sol verger. Annales Agronomiques 16 (5): 553–577
12. Monteny B and Eldin M (1977) Données agroclimatiques recueillis à la station ORSTOM d'Adiopodoumé: 1948–1976. ORSTOM, Adiopodoumé, RCI. pp. 24
13. Olsen JS (1963) Energy storage and the balance of producers and decomposers in ecological systems. Ecology 44: 322–332
14. Read MD, Kang BT and Wilson GF (1985) Use of *Leucaena leucocephala* (Lam.) de Wit leaves as a nitrogen source for crop production. Fertilizer Research 8(2): 107–116
15. Russell EW (1973) Soil conditions and plant growth. ELBS, Longman. 10th edition. pp. 849
16. Snedecor GW and Cochran WG (1973) Statistical methods. Iowa University Press. 6th edition. pp. 593
17. Yamoah CF, Agboola AA and Mulongoy K (1986) Decomposition, nitrogen release and weed control by prunings of selected alley-cropping shrubs. Agroforestry Systems 4: 239–246

Agroforestry Systems 6: 137–145 (1988)

5

The performance of the leaf mulches of *Leucaena leucocephala*, *Flemingia macrophylla* and *Gliricidia sepium* in weed control*

ARNOUD BUDELMAN
Centre Neerlandais, B.P. V51, Abidjan, Ivory Coast

Key words: *Leucaena leucocephala*, *Gliricidia sepium*, *Flemingia macrophylla*, leaf mulches, weed control

Abstract. The performance of the leaf mulches of *Leucaena leucocephala*, *Gliricidia sepium* and *Flemingia macrophylla* in weed control has been tested in two trials. The length of the period during which a mulch layer yields significantly less weedbiomass compared to the control plots is called the 'effective life-span' of the mulch.

Of the three mulch materials only that of *F. macrophylla* shows promise in retarding weed development.

In the second trial *F. macrophylla* leaf mulch was applied at rates of 3, 6 and 9 tons dry matter per ha. The effective lifespan of a mulch layer of 3 tons is between 12 and 13 weeks. The treatments 6 and 9 tons have effective life-spans of over 14 weeks.

For moderate quantities (up to 5 tons of dry leaf mulch per ha) the effective life-span is estimated at about a 100 days.

The value of mulching in weed control is limited to the control of weed species that multiply by seed. Regrowth originating from roots or stumps from former vegetation is unlikely to be checked by a mulch layer.

Introduction

In the past few years, the interest of the scientific community in leguminous shrubs and trees capable of fixing nitrogen has grown significantly. One explanation is the increasing awareness that nitrogen fertilisers have become prohibitively expensive, particularly in the context of food crop cultivation. Moreover, in many parts of Africa delivery systems for farming inputs are ineffective or simply non-existent.

Besides being valuable as a cheap and accessible source of nitrogen, the shrubs and trees, when selected correctly, have other purposes as well when integrated in cropping systems.

* The subject of this paper is part of a broader study, presently conducted in Southern Ivory Coast, on the production, agronomical value and use in alley-cropping systems of the species mentioned in the title of the paper. A final report is scheduled for publication in 1988.

Three woody leguminous species are currently studied as living support systems for yam cultivation; *Gliricidia sepium*, *Leucaena leucocephala* and *Flemingia macrophylla* (syn. F. congesta) [6].

Of the species in observation *G. sepium* is used by farmers in Western Nigeria in the role as support for yam plants [1, 8] and is found as a planted fallow [8, 13]. In the latter function the tree is known to be effective against invading Speargrass (*Imperata cylindrica*) [2, 8]. A further development is the design of integrated cropping systems matching trees and annual crops such as alley-cropping [11].

All three species are accepted by ruminants as fodder, either in cut-and-carry feeding systems, or as plants suitable for dry season browsing [4, 7, 16].

The latest perspective is to link crop and livestock production in systems centered around the use of multi-purpose trees and shrubs [16]. Such mixed production systems may eventually become part of a strategy to replace shifting cultivation practices.

Both *Gliricidia* and *Flemingia* have been used to provide shade in coffee and cocoa research in francophone Africa [5, 17], while *F. macrophylla* once was tested in a screening of cover crops suitable in rubber plantations in Ghana [3].

However, apart from *Leucaena leucocephala* [11, 15] few detailed studies exist on the agronomical value of the species in systems including annual food crops.

One of the possible functions of perennials in cultivation systems is to regularly apply the biomass produced as a mulch layer in order to control weed growth. For that purpose the leaf mulch must have a certain persistence. Especially when high rainfall is combined with high temperatures, the decomposition of organic material is accelerated, and, consequently, the effective life of the mulches is shortened.

In literature most of the studies refer to the effects of mulches on soil temperature and moisture content [cf. 12] or on the soil fertility status [cf. 15, 18]. Only a few studies exist that provide factual information on the effects of mulch on the development of weeds [9, 19].

This paper describes the results from two trials comparing the leaf mulches of *Leucaena leucocephala*, *Flemingia macrophylla* and *Gliricidia sepium* for their effectiveness to reduce weed growth.

Site of the trials, materials and methods

The trials were conducted on the experimental farm on the ORSTOM research station near Abidjan, Ivory coast. The farm is located on a well-drained sandy soil (clay content < 15%; pH 4.7; CEC 3–5 Meq.; strongly desaturated, typically < 15%).

The annual rainfall is about 2,000 mm. The trials were carried out during the long rains season. In the period the average daily rainfall exceeded the evapotranspiration. The results as a consequence must be considered as reflecting a *wet regime* situation.

Seeds from *Leucaena leucocephala* were collected from the 'Hawaiian' type locally found in roadside vegetations. Seeds from *Flemingia macrophylla* and cuttings from *Gliricidia sepium* were acquired from the 'Institut de Recherche du caféier et du Cacao' at Divo, Central-South Ivory Coast, where both plant species are used as cover crops and shade trees.

The mulch used in the weed control trials came from productivity trials involving all three species. At the time when the first trial took place the plantations were two years old. In the second trial the material came from a *F. macrophylla* plantation aged three years.

Before the mulch was placed in the field, the trial area was completely clean-weeded. Although the trial was situated on a flat soil, the mulch layers were secured against eventual run-off using bamboo frames that marked the experimental units of 2 by 2 m (first trial) or 2 by 3 m (second trial). The mulch application rates were 5 ton dry material for each of the species in the first trial, and 3, 6, and 9 ton dry matter of *F. macrophylla* during the second trial.

The dry matter contents, half-life values and leaf surface to weight ratios of the leaf mulches are given in Table 1.

In the first trial a Latin Square was used as experimental design. This design was chosen with the aim to control variation in individual weed species emergence in two directions. Samples taken measured 50 × 50 cm.

Since weed emergence at the species level was heterogenous and often patchy, the original Latin Square design was abandoned in favour of a

Table 1. Dry matter content, decomposition and leaf surface to weight ratio of the leaf mulch materials used (note: the 'half-life' value represents the number of days to reach a 50% loss of the mulch material)

Species	Dry matter content		'Half-life' value	Leaf surface to weight ratio
	2 yr. old trees	3 yr. old trees		
L. leucocephala	26.5%	–	30.7	193.7
F. macrophylla	28.6%	27.6%	53.4	214.5
G. sepium	21.3%	–	21.9 days	180.2 sq.cm/g DM

(source: 6)

Randomised Block design, with sampling squares that measured 71 × 71 cm.

Per sampling date one sample was taken from each of the experimental units. Both trial designs included four replications.

Results and discussion

In Figure 1 the results of the first trial are given. It is evident that *L. leucocephala* is practically without value when its mulch is applied for reasons of weed control. Almost all weed biomass-values observed in the *L. leucocephala*-covered plots fall within the limits of non-significant differences. Better results are obtained with *G. sepium* and *F. macrophylla*. The latter is certainly able to suppress weed growth to a considerable extent.

From Figure 1 and Table 1 it can be concluded that the differences in behaviour of the mulches of *G. sepium*, *F. macrophylla* and *L. leucocephala*

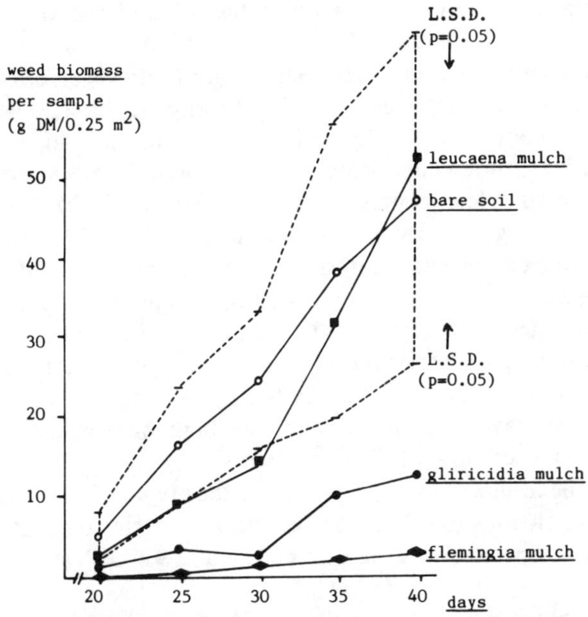

Fig. 1. The effect of the mulches of *Gliricidia sepium*, *Flemingia macrophylla* and *Leucaena leucocephala* on the development of the weed biomass as compared to the weed biomass collected on an un-mulched soil (note: the area in Fig. 1 within the dotted lines represents the weed biomass quantities that do not significantly differ from the weed biomass found on the un-treated experimental units)

regarding weed suppression can not be attributed to mere differences in the dry matter content or the area to weight ratio of the leaves. Generally speaking, leaves with a low area to weight ratio are relatively resistant to decay [10].

An explanation for the differences observed is more likely to be found in the 'morphology' of the mulch layers once dried. In the case of *L. leucocephala* the compound leaf consists of many very small leaflets (0.7 to 1.0 cm), which almost immediately fall from the rachis when drying starts. The structure of the mulch layer that results resembles strongly that of a collection of dried tea leaves. Unprotected, *L. leucocephala* mulch is easily blown away, or transported by surface run-off.

In contrast *F. macrophylla* has large leaflets, of 10 to 20 cm long, that stay firmly attached to the petiole. The leaflets have a strong tendency to curl when drying, creating a relatively voluminous, coherent layer in which considerable quantities of air can be trapped.

The structure of the mulch layer also explains the differences found in the decomposition rates; see Table 1. The compact structure of the mulch layer of *L. leucocephala* has more contact with the soil underneath, compared to the loosely packed *F. macrophylla* mulch.

G. sepium takes a position in between; Figure 1. However, considering the quick decomposition of the leaf material shown in Table 1, the value of its mulch in weed control will be limited. This conclusion is supported by the results from field studies conducted at IITA, where *G. sepium* prunings were found to have completely vanished within a period of 120 days [19].

The resistance against decomposition of the mulch of *F. macrophylla* as compared to the other mulch materials can further be explained when taking a look at the level of its chemical composition. *F. macrophylla* leaves, subject to an in-vitro digestibility test, were found to decompose twice as slow as the leaves of the other two species involved [6]. This suggests that the percentage of dry matter is less important than its quality. A likely interpretation is that *F. macrophylla* leaves contain a relatively high proportion of lignins, a chemical substance known to break down slowly.

Based on the results of the first trial series the mulch of *F. macrophylla* was selected for further testing. In the second trial increasing quantities of mulch were applied, respectively at the rate of 3, 6 and 9 tons leaf dry matter per ha.

Figure 2 shows that all three treatments have a marked effect on the development of weed growth. Even in the case of the lowest amount of 3 tons dry leaf matter per ha applied, weed biomass development is significantly retarded. The *effective life-span* (= the period during which the weed biomass collected from a mulched plot differs significantly from that harves-

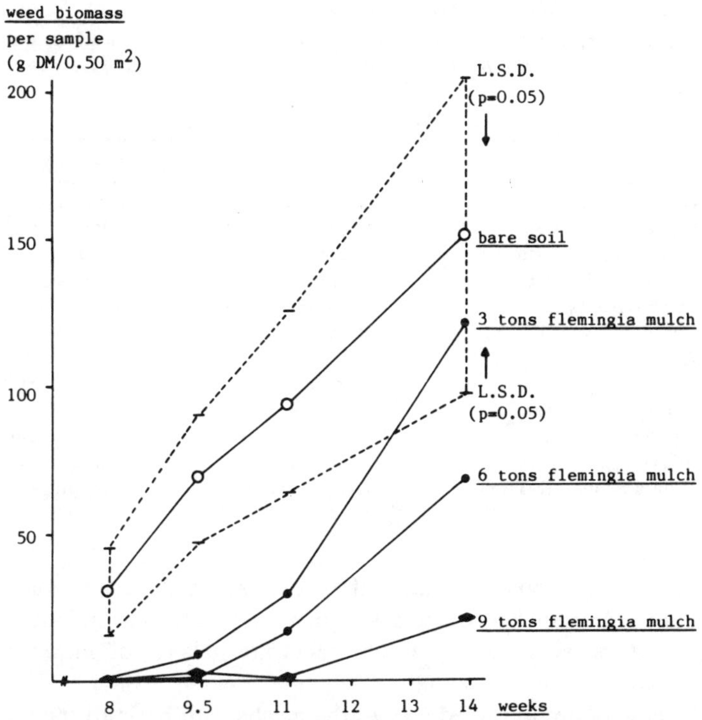

Fig. 2. The effect of increasing quantities of mulch of *Flemingia macrophylla* on the development of the weed biomass as compared to the weed biomass collected on an un-mulched soil (note: the area in Fig. 2 within the dotted lines represents the weed biomass quantities that do not significantly differ from the weed biomass found on the un-treated experimental units)

ted from the untreated plots) of a mulch layer of 3 ton is between 12 and 13 weeks, or the time it takes a maize crop to mature.

The higher application rates show longer effective life-spans simply as a result of the larger amounts of material.

However, a *F. macrophylla* mulch layer loses its effectiveness when regularly trod on, especially when the material is not wetted by rains. Drying out causes the material to become brittle, after which it disintegrates easily.

In Table 2 the most important weed species present in the early regrowth stage after clean-weeding are listed.

It should be emphasised that all the weed species enumerated in Table 2 are annuals, reproduced from seeds. *Cyperus rotundus* is the sole exception, since this species is propagated by small, connected tubers. In the latter case as well, the young plants that emerge are notably hindered in their development by the mulch layer.

Table 2. Weed species present during the second trial
(notes: — species represented with less than 5 individuals in the entire trial are not listed,
— frequent = more than five individuals, but not present in all of a series of samples taken,
— dominant = species present in all of a series of samples)

Species	Family	Remarks
Cyperus rotundus L.	Cyperaceae	dominant
Mariscus alternifolius Vahl	Cyperaceae	frequent, later stage
Eleusine indica (L.) Gaertn.	Poaceae	frequent
Eragrostis ciliaris (L.) R.Br.	Poaceae	frequent
Ageratum conyzoides L.	Asteraceae	dominant
Borreria latifolia (Aubl.) K. Schum.	Rubiaceae	dominant
Oldenlandia corymbosa L.	Rubiaceae	frequent, early stage
Mollugo cerviana (L.) Ser.	Molluginaceae	dominant, early stage
Mollugo nudicaulis Lam.	Molluginaceae	frequent, early stage
Celosia trigyna L.	Amaranthaceae	frequent, later stage
Cleome ciliata Schum. & Thon.	Capparidaceae	frequent, later stage

There is however no doubt that leaf mulches are ineffective against growth of root and stump suckers of trees and shrubs left in the ground after cutting a vegetation, since these can easily penetrate a layer of loosely packed organic matter.

In the case of annual weed species the mechanism basic to the success of the leaf mulch is either prevention of germination (in cases where light acts as a trigger in the germination process), or stunting the early development of the small seedlings.

Allelopathy is a third potential factor that may explain the role of mulches in retarding weed development. However, data presented here do not warrant conclusions in that direction.

Conclusion

Of the three leaf mulches tested only *F. macrophylla* shows promise when it comes to retarding weed development. The slow decomposition of pruned material [6, 19] and its relatively easy control as an alley- or covercrop makes the species of interest in a variety of cropping systems. In rubber plantations in Ghana the number of weeding rounds could be reduced from six to two times a year using *F. macrophylla* as a covercrop [3]. As a covercrop and regularly cut applied as mulch *F. macrophylla* considerably reduced weed infestation in coffee plantations [17].

Neither *G. sepium* nor *L. leucocephala* are suitable in a role in weed-control, since both simply decompose too fast to have a significant, lasting effect [6, 14, 19].

The data presented show that the success of mulches in weedcontrol depends at least partly on intrinsic, or material-specific factors, such as its behaviour when drying and the chemical composition of its dry matter. *Flemingia macrophylla* is an example of a source of material that combines a notable durability with a favourable ratio between volume and weight. Compared to other sources [9] one needs relatively small quantities to achieve complete soil cover.

It is likely that other factors can be mobilized as well that improve effect and/or durability of mulches. In that respect one can think of material with allelopathic properties, material high in tannin content, or to test plant species with sclerophyllous leaves, such as *Ficus benjamina*, that are tough and resistant to decay during the first stage of decomposition.

The suggestion here is to make an inventory of plant species that have one or more of the properties mentioned and to see whether these species can be integrated to serve in low external input cropping systems.

References

1. Agboola AA (1982) Gliricidia sepium: A possible means to sustained cropping in Western Nigeria. In: MacDonald LH (Ed) Agro-forestry in the African humid tropics. Tokyo, Japan (ISBN 92-808-0364-6)
2. Aken'ova ME and Atta-Krah AN (1986) Control of Speargrass in an alley-cropping fallow. NFTA Res Rep 4: 27–28
3. Anon (1964) Ann Rep Agric Res St Kade (Ghana) 1963–64. Memorandum no 3. Univ Ghana
4. Asare EO, Shehu Y and Agishi EA (1984) Preliminary studies on indigenous species for dry season grazing in the Northern Guinea Savanna Zone of Nigeria. Trop Grassl 18(3): 148–152
5. Bouharmont P (1979) L'utilisation des plantes de couverture et du paillage dans la culture du caféier 'Arabica' au Cameroun. Café, Cacao, Thé, 23(2): 75–100
6. Budelman A (in preparation). Alternatives to shifting cultivation: The use of woody leguminous perennials in yam cultivation. Department of Tropical Crop Husbandry, Agric Univ Wageningen, The Netherlands
7. Chadhokar PA and Sivasupiramaniam S (1983) Gliricidia leaves as protein supplement to paddy straw in growing cross-bred heifers. Indian J Anim Sc 53(2): 120–125
8. Clayton WD (1958) Secondary vegetation and the transition to savanna near Ibadan, Nigeria J Ecol 46(2): 217–238
9. IITA (1982) International Institute of Tropical Agriculture Annual Report 1982, Ibadan, Nigeria (vide p. 39)
10. Jordan CF (1985) Nutrient cycling in tropical forest ecosystems. Principles and their application in management and conservation. John Wiley and Sons, Chichester, UK (vide p. 12)

11. Kang BT, Wilson GF and Lawson TL (1984) La culture en couloirs. Un substitut d'avenir à la culture itinérante. Tech Bull, IITA, Nigeria
12. Lal R (1974) Soil temperature, soil moisture and maize yield from mulched and unmulched tropical soils. Plant and Soil 40: 129–143
13. Okigbo BN (1983) Plants and agroforestry in land use systems of West Africa. In: Huxley PA, 1983. Plant Research and Agro-forestry, ICRAF, Nairobi, Kenya
14. Read MD (1982) (quoted by KANG et al 1984 [11] on p 17; figure 12)
15. Read MD, Kang BT and Wilson GF (1985) Use of Leucaena leucocephala (Lam.) de Wit leaves as a nitrogen source for crop production. Fertilizer Research 8: 107–116
16. Sumberg JE (1984) Alley-farming in the humid zone: Linking crop and livestock production. ILCA Bull 18: 2–6
17. Verlière G (1966) Valeur fertilisante de deux plantes utilisées dans les essais de paillage du caféier: Tithonia diversifolia et Flemingia congesta. Café, Cacao, Thé 10(3): 228–235
18. Wade MK and Sanchez PA (1983) Mulching and green manure applications for continuous crop production in the Amazon Basin. Agronomy Journal 75: 39–45
19. Yamoah CF, Agboola AA and Mulongoy K (1986) Decomposition, nitrogen release and weed control by prunings of selected alley-cropping shrubs. Agroforestry Systems 4: 239–246

Agroforestry Systems 8: 53–66, 1989.

6
The performance of selected leaf mulches in temperature reduction and moisture conservation in the upper soil stratum

ARNOUD BUDELMAN
Centre Néerlandais, Abidjan, Côte d'Ivoire, Department of Tropical Crop Science, Agricultural University, Wageningen, The Netherlands
Present address: Royal Tropical Institute. Mauritskade 63, 1092 AD Amsterdam, The Netherlands

Key words: leaf mulches, soil temperature reduction, soil moisture conservation, *Leucaena leucocephala, Gliricidia sepium, Flemingia macrophylla*

Abstract. This paper reports on the effect of the leaf mulches of *Leucaena leucocephala, Gliricidia sepium* and *Flemingia macrophylla* on moisture content and temperature in the first 5 cm of the soil. The mulches were applied at a standard quantity of 5,000 kg ha^{-1} DM.

In order to characterise a mulch material two parameters are distinguished; the *initial impact* (I_i) and the *effective lifetime* (T_e) of a mulch material. I is expressed in terms of percentage surplus moisture or degrees Celsius average temperature reduction. T quantifies the duration of the effect. Of the three mulch materials that of *Flemingia macrophylla* performs best in terms of moisture retention and lowering soil temperatures as well as in terms of longevity of the effect. *Leucaena leucocephala* mulch shows the smallest impact, over the shortest period.

1. Introduction

Appropriate definitions of a mulch have been given by Mellaart [1975] and Stigter [1984]. In recent years scientific interest in mulching as a low-cost technique for use in smallholding cultivation in the tropics has returned. One reason is that it becomes increasingly clear from research that mulches promise a range of potential benefits in tropical cropping systems; erosion- and weed control, soil moisture conservation, soil temperature management and addition of nutrients. None of these benefits is particularly spectacular in terms of its effect on crop yield, but combined they provide ecologically more sound conditions for crop growth.

A second reason explaining the growing interest among scientists concerned with tropical cropping system development is that mulching requires no other inputs than land to grow and labour to collect mulch material. In regions with a poorly developed infrastructure, mulching, as part of a so-called *low external input agricultural technology*-package [cf. Sanchez & Salinas, 1981], may provide farmers with accessible, economically viable and ecologically sustainable alternatives to shifting cultivation.

Literature shows that factual knowledge on the use and effects of mulches is available [2, 5, 6, 7, 14, 15], but understanding remains incomplete. Information on the rate of decomposition of *organic* mulches, for example, is scarce [3, 9], while this particular aspect is important since it determines e.g. the release of nutrients from the material and the effective life-time.

This paper quantifies the effect of the leaf mulches of *Leucaena leucocephala, Gliricidia sepium* and *Flemingia macrophylla* on moisture and temperature in a sandy, well-drained ferralitic soil in Ivory coast [1]. The study is a follow-up of earlier papers on the performance of the mulches in weed control [2], the nutrient composition [4] and on the decomposition characteristics of the leaf materials [3].

In the tropics mulching is a well-suited technique to lowering average soil temperatures and to conserving water. Both aspects are particularly important in the early stages of crop growth; drought may cause the death of seedlings, while excessively high temperatures result in low emergence rates, stunted shoot- and insufficient root development [6]. From these observations it can be concluded that knowledge on the impact of a certain mulch material at a given time in a cropping cycle is most relevant if mulches are

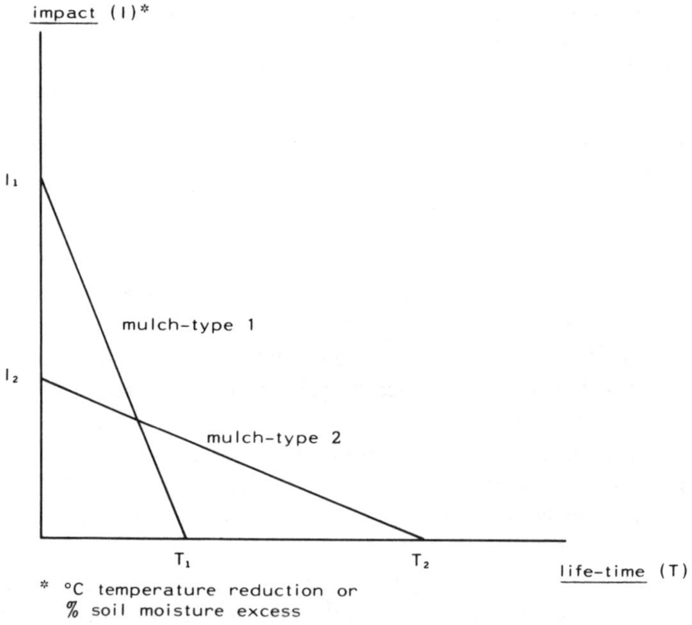

Fig. 1. Behaviour of mulches as a function of material-specific factors (simplified presentation).

expected to alleviate such problems. It is likely that mulch materials will differ in their impact (*I*) as a function of material-specific properties. The parameter (*I*) is measured in terms of degrees temperature reduction or percentage soil moisture surplus.

A second important variable is the actual duration of the effect; the period there is an agronomically significant difference between mulched and unmulched fields. This parameter, called the *effective life-time* (T_e), depends on the resistence which the mulch material shows against decomposition. Obviously, the scope of the effects also depends on the depth in the soil where the effect is measured.

The simplified model presented in Figure 1, illustrating the effect of mulches in time, can in principle be adapted if the reduction of the effect takes a non-linear shape, e.g. an exponential decrease. In that case, fitting the experimental data to relevant mathematical functions will provide an additional (set of) variable(s) characteristic of such a mulch material. Other approaches may be used as well [9].

In Figure 1 both parameters are presented for two different hypothetical mulch materials. A mulch material that answers 'type 1' would be e.g. a layer of leaves with a low C/N ratio. This would guarantee rapid turn-over by soil-dwelling insects that use the mulch as a source of food and dissimilation of the organic matter by micro-organisms. However, the initial impact on soil temperature and moisture can be considerable.

A thin layer of wood chips (high C/N ratio) would fit a 'type 2' mulch material. Its impact on soil temperature is small but long-lasting when not disturbed. Obviously, the comparison is correct only when using equal quantities of dry matter per unit area.

That condition brings us to one of the limitations of this study, since the mulch materials are compared and parameters determined at a standard level of application of 5000 kg ha^{-1} DM only, while it is likely that the outcome of the parameters investigated will vary according to the quantity of material used.

2. Location and climate

The trial was conducted on the experimental farm of the ORSTOM research station near Abidjan, Ivory Coast. In Table 1 some relevant physical and chemical parameters are specified.

The trial was carried out during *a dry period*: Over the complete period of the trial (60 days) a total of 59.5 mm rainfall was recorded, including two separate showers of respectively 20 and 22 mm. The potential evapo-trans-

Table 1. Selected physical and chemical parameters of the soil of the trial area; 0–20 cm (abbreviated after Siemonsma, 1982).

% sand 50–2000 μm	% clay + loam < 50 μm	pH H_2O	O.M. %	C ‰	N ‰	P* ‰	CEC** Meq.	base saturation %
84.5	14.1	4.7	1.06	6.11	0.59	0.24	3.27	13.6

* total P (HNO_3): **per 100 g soil

piration during the trial period totalled 210 mm. The average maximum air temperature was 31.2 °C.

Climatological data came from the ORSTOM department of bioclimatology. The meteorological field-station was located c. 500 m from the trial area. For further information on climate and soil see ref. 1.

3. Materials and methods

Materials

Seeds from *Leucaena leucocephala* (Lam.) de Wit were collected from the 'Hawaiian' type locally found naturalised in roadside vegetations. Seeds from *Flemingia macrophylla* (Willd.) Merr. and cuttings from *Gliricidia sepium* (Jacq.) Walp. were acquired from the Coffee and Cacao Research Station at Divo. Central South Ivory Coast, where both plant species are used as cover-crops.

The leaf* mulch used in this experiment was collected from productivity trials involving the three woody perennials [1].

Trial design and implementation

The trial was located in an open, unshaded area, surrounded by blocks with cover-crops. Adjacent blocks of vegetation reduced the effect of wind significantly. The minimum distance between the plants in the blocks and the border of the experimental units was 2.5 m. Bamboo poles were used to mark 2 by 3 m experimental units that were situated next to each other in a 4 by 4 randomised block design: 4 treatments, 4 replications. Including the corridors between the blocks (spaced 1 m apart) the total trial area measured 8 by 15 m.

* A leaf is defined as the complete compound leaf, including *pulvinus, petiolus, rhachis* and *foliola*.

The treatments were i) uncovered soil (control), ii) soil covered with leaf mulch of *Leucaena leucocephala*, iii) – leaf mulch of *Gliricidia sepium* and lastly, iv) – leaf mulch of *Flemingia macrophylla*.

The mulch was applied as fresh leaves at a rate of 5000 kg ha^{-1} DM, and evenly distributed in each of the clean-weeded experimental units. The bamboo frames were meant as a prevention against dislocation of the mulch layers. A light cover of stripped *Leucaena leucocephala* branches (3 to 4 per experimental unit) was applied as an extra precaution. Emerging weed seedlings were removed daily. This guaranteed a minimum disturbance of the mulch layers.

Measurements and sample treatment

Temperature was read at 15.00 h, at 5 cm depth. The time corresponds with the approximate moment the temperature in the upper soil strata under all treatments reached its peak-value. After reading the temperature soil samples were taken.

Temperature measurements and soil sampling always took place at exactly the same relative position in each of the experimental units in order to limit variation. The sampling position was shifted in a random manner.

During each sampling round, per experimental unit two temperature readings were made. This was done with the help of two calibrated soil-thermometers. The average was noted as the reading for a particular experimental unit.

Soil moisture was determined gravimetrically. Soil sampling was done with a steel cylinder that was 5 cm long with a diameter of 50.5 mm, and a content of 100 cm^3. The cylinder permitted very precise sampling over the first 5 cm in the soil.

The samples were then immediately put in closed aluminium containers of known weight. After weighing the 'wet' soil, the containers were put into a large ventilated box in which strong lamps provided a continuous temperature of c. 50 °C. A fan at one side of the box kept the air moving. Samples were regularly stirred and air-dried during 24 hours. Dry weight was then determined.

Calculations and statistics

Observations took place over a period of 60 days. A total of 10 series of measurements were collected. Each series of readings was subjected to an analysis of variance that gave averages for the four treatments, F-values for treatment and block effects and a value for the Least Significant Difference

(L.S.D.). The L.S.D.-value was calculated according to the *Studentised Range*-test, which is identical to TUKEY's *Multiple Range*-test, and permits comparison among means of individual treatments [12; p. 271). In order to limit variation in data and to facilitate comparison between series the following procedural steps or modifications were made;

a. no soil sampling was done within 72 h after a shower.
b. temperature measurements only took place on clear, sunny days in order to maximise the potential of the mulch layers to reduce soil temperature.
c. instead of using the actual average readings in statistical analysis the average reading from the control was subtracted from that found for the treated plots. The value thus acquired for each of the treatments differed significantly from the control situation as long as it exceeded the L.S.D. value for the particular series of samples taken (Figures 2 and 3).

The data thus derived were used in regression analysis, with the time in days on the X-axis. With the help of the regression equations a value for the *initial impact* (I_i) was calculated at t = 10 days, either as a percentage surplus moisture or as °C temperature reduction. The first 10 days are considered as a period necessary to allow the mulch layer to settle.

The second parameter that matters in this characterisation of mulch materials depends on the *regression coefficient* which determines the duration of the period that mulches can be considered as effective in terms of water conservation and reduction of soil temperature; the *effective life-time* (T_e).

The average L.S.D. at $P = 0.05$ is used as the lower limit for which there is a statistically significant difference between treatment and control; Figures 2 & 3.

pF curve

In order to put the scope of moisture conservation caused by the three mulch materials into perspective, a *pF-curve* for the soil *stratum* 0–5 cm was determined by the ORSTOM central laboratory at Adiopodoumé, Ivory Coast. The curve was constructed from 8 soil samples at four different pF values. The best fitting curve was $Y = 30.3e^{-0.5x}$ with $r^2 = 0.92$, in which X represents the pF-values and Y the related soil moisture content in percentages. As value for the upper limit of moisture in the soil ATTERBERG's limit was determined. As lower limit soil moisture at pF = 4.2 was experimentally measured. The range thus found ran from 3.7 to 14.2% for this sandy type of soil.

4. Results

4.1 The structure and colour of the dried mulch materials

Each of the leaf mulches shows a typical structure and colour once dried. Structure and colour stem from leaf-morphological and chemical aspects that are specific for each of the plant species.

Leucaena leucocephala leaves are bi-pinnate, with many, very small (0.7–1.0 cm) leaflets. During drying the leaflets fall from the *rhachis*, and the mulch layer consequently consists of a loose collection of small particles closely resembling dried tea leaves. Obviously, the lack of coherance in the mulch of *Leucaena leucocephala* increases the chance of being blown away or taken by surface run-off. From the start the dried mulch layer is relatively thin; at 5000 kg ha^{-1} DM a few cm thick.

The mulch of *Leucaena leucocephala* is distinctly deep dark brown in colour, and provides by far the darkest surface of the three mulches studied.

In contrast, *Flemingia macrophylla* leaves have only three large leaflets, of 10 to 20 cm long, that stay firmly attached to the petiole. During drying the leaflets curl, creating a relatively voluminous, coherent layer (initially at least 10 cm thick). A mulch layer of *Flemingia macrophylla* resembles, both in structure and colour, a mulch of beech-leaves. Of the three mulches *Flemingia macrophylla* has the lightest colour.

The pinnate leaves of *Gliricidia sepium* provide a mulch type that takes an intermediate position between the other two in terms of structure and colour. Here, the leaflets have a tendency to fall from the *rhachis* as well. But they are much larger (up to 5 cm) than the leaflets of *Leucaena leucocephala*, and layers of this particular mulch appear more voluminous. The colour is greyish-green.

4.2 Average temperature and moisture content under the mulches

In Table 2 observations on soil temperature at 5 cm depth and soil moisture content in this *stratum* are presented in absolute average values over the complete period of the trial.

Using *Student's* t-test all averages turn out to be significantly different from each other at the P = 99.9% confidence level, except for the difference in humidity under the mulches of *Gliricidia sepium* and *Flemingia macrophylla*, which is significantly different at the 95% level. This result shows that over time there is remarkably little variation in the observations and that each of the mulch materials has a specific soil temperature and soil humidity range.

Table 2. Average temperature and soil moisture content over a period of 60 days as a function of three different mulch materials (initial rate of application 5 ton ha^{-1} DM).

Treatment/ mulch material	Nr. of observations at 15.00 h	Average temperature at 5 cm	Average % soil moisture over 0–5 cm
Unmulched soil	40	37.1 °C	4.8%
Leucaena leucocephala	40	34.2 (−2.9)*	7.1 (+2.3)*
Gliricidia sepium	40	32.5 (−4.6)	8.7 (+3.9)
Flemingia macrophylla	40	30.5 (−6.6)	9.4 (+4.6)
L.S.D. value**		1.20 °C	1.84%
standard deviation		±0.47 °C	±0.39%

* between brackets the difference relative to an unmulched soil
**average and standard deviation of the 10 series of observations

Of the three mulches *Leucaena leucocephala* shows the weakest performance.

From the point of view of moisture conservation and temperature reduction *Flemingia macrophylla* provides a valuable mulch material. Temperature reduction is considerable, while the presence of the mulch almost doubles the average soil moisture content over the soil *stratum* sampled.

The effect of the mulch of *Gliricidia sepium* takes a position in between that of the two other mulch types. This sequence in performance has already been noted in an earlier publication in which the three mulches were tested in a weed control trial [2].

4.3 Initial impact and effective life-time of the mulches

In Figures 2 and 3 the influence of the mulch types on the soil temperature and soil moisture content is related to the age of the material. Fitting the data to a linear regression line yields high levels of statistical significance.

The general conclusion is that, in the course of time, the effect of the mulches on the soil temperature and soil moisture diminishes. This is caused by the fact that mulches decompose resulting in decreasing quantities of material that influence the energy balance at the soil-air interface.

In the case of soil temperature the rate by which the effect decreases is about equal for the three materials. This rate, the *regression coefficient* in the equation, can be estimated at 0.12 °C day^{-1}. Where the mulches differ most is in their respective initial impact on the soil temperature; Table 3. The mulch of *Flemingia macrophylla* is much more effective in lowering the soil temperature compared to the other two types of material.

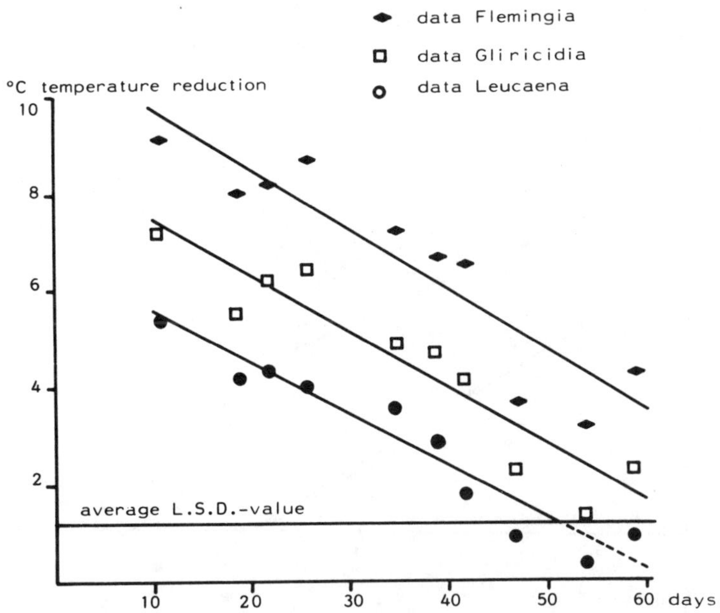

Fig. 2. Temperature reduction as a function of the age of the mulch layers.

		r^2	significance
Leucaena leucocephala	Y = −0.108X + 6.67	0.94	P < 0.001
Gliricidia sepium	Y = −0.116X + 8.61	0.88	P < 0.001
Flemingia macrophylla	Y = −0.124X + 11.00	0.84	P < 0.001

for statistical significance of r^2 see ref. 12: p. 184

The regression lines that describe the diminishing effect of the mulch material on soil moisture show a similar pattern as in the case of the soil temperature. Although still sufficient in terms of statistical significance the fit is lower than in the case of the temperature series. This may result from differences in the size of the leaves of the species tested. The small leaves of *Leucaena leucocephala* permit rain water or dew to penetrate the soil in a more homogenous manner as compared to the large leaves of *Flemingia macrophylla*. As a consequence the variation in the readings expressed in the lower r^2-values (see legends Figures 2 and 3) is considerably greater in the latter case.

The mulches of *Gliricidia sepium* and *Flemingia macrophylla* lose their capacity to influence soil moisture at an equal pace, while the greater

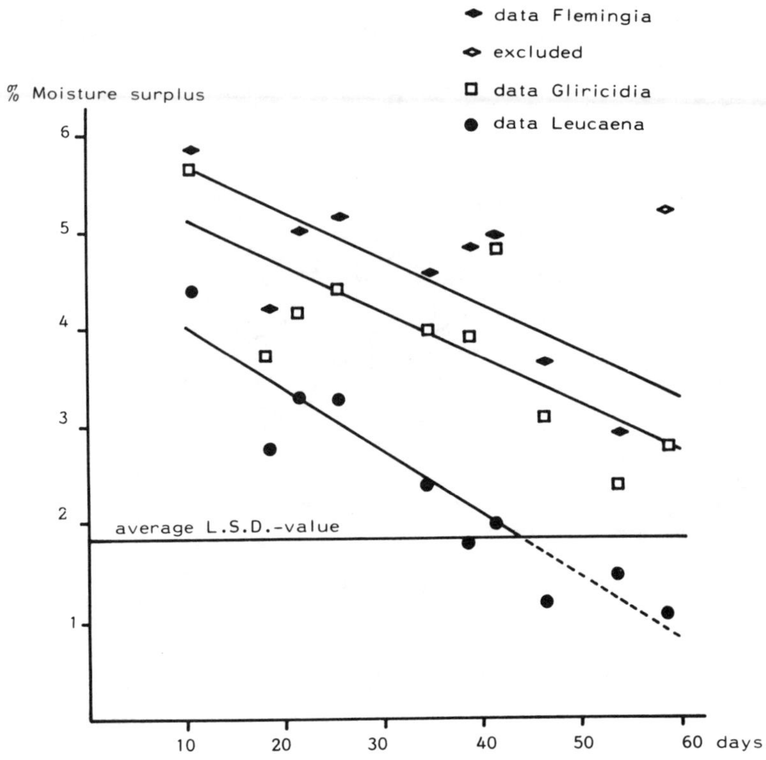

Fig. 3.: Soil moisture surplus as a function of the age of the mulch layers.

		r^2	significance
Leucaena leucocephala	$Y = -0.063X + 4.59$	0.88	$P < 0.001$
Gliricidia sepium	$Y = -0.047X + 5.57$	0.58	$P < 0.025$
Flemingia macrophylla	$Y = -0.047X + 6.11$	0.57	$P < 0.025$

for statistical significance of r^2 see ref. 12: p. 184

regression coefficient of *Leucaena leucocephala* indicates a quicker end to its role in conserving soil moisture. With the help of the regression equations from Figures 2 and 3 the two parameters, *initial impact* (I_i) and *effective lifetime* (T_e), can be estimated. The initial impact of the mulches is calculated for 10 days old mulch layers, rather than for 'zero'-time. The rationale is that new equilibria have to be established under the mulch layers, particularly with respect to soil moisture. Substituting 10 for X in the two series of regression equations gives the values for the *initial impact*.

Table 3. Initial impact and effective life-time (calculation mode: 'initial impact' at X = 10 days; 'life-time' at Y = 1.20 (av. L.S.D. temperature) or Y = 1.84 (av. L.S.D. moisture) see Table 2 and Figures 2 & 3)

Mulch type	Initial impact*	Effective life-time
Temperature		
Leucaena leucocephala	−5.6 °C	51 days
Gliricidia sepium	−7.5 °C	c. 65 days
Flemingia macrophylla	−9.8 °C	> 70 days
Moisture		
Leucaena leucocephala	+4.0%	44 days
Gliricidia sepium	+5.1%	c. 80 days
Flemingia macrophylla	+5.6%	> 90 days

* relative to the situation under unmulched soil

The *effective life-time* (T_e) is calculated with the same regression equations but then by substituting Y. The average value of the Least Significant Difference (L.S.D.) was chosen instead of Y = 0 (Table 3; Figures 2 & 3) as a relevant value to determine the end of the *effective life-time*. The average L.S.D.-value takes into account the part of the effect that can be attributed to error or unexplained variation in the series of observations. In the calculations a confidence level of 95% is assumed to be sufficiently strict. The *effective life*-time of the mulch of *Leucaena leucocephala* can be estimated with a reasonable degree of precision (Figures 2 & 3, Table 3) and is just under 50 days.

Of the remaining mulch materials that of *Flemingia macrophylla* shows the greatest longevity. Regarding both effects, only estimated effective life-times can be given. Over the period of observation the regression line never meets the lower limit for significant difference. This is also true for the mulch of *Gliricidia sepium*.

5. Discussion

As the results of this study show, the three mulch materials have a significant potential to influence temperature and moisture in the upper soil *stratum* (0–5 cm). Reduced temperatures and increased, regular moisture availability in the upper soil strata is important in relation to early stages of crop development (germinating seeds, sprouting in vegetative propagation), superficially rooting crop species and nitrogen fixation in root nodules. Apart from the enhanced sprouting rate of yam setts reported elsewhere [5] it was found that root-nodulation in stands of *Gliricidia sepium* was most intensive on the interface between the mulch layer and the soil surface. Both observations typify the use of mulches as a tool in micro-climate manipulation inducing better growth conditions [see also ref. 6, 7, 13].

Potentially, differences in performance of the mulch materials in terms of soil temperature reduction, moisture conservation and longevity are a function of i) the structure and relative volume, ii) the colour of the dried mulch layer, and iii) the chemical composition of the leaf biomass.

Evidently, initial differences in the mulches' capacity to reduce soil temperature and to conserve water are related to structure and volume of the material once it is dried. The voluminous layer formed by the mulch of *Flemingia macrophylla* traps considerable quantities of air which form a barrier to heat transport and evaporation of soil moisture. This effect is much less in the case of the other two mulch materials because of the thinness of the layers.

A second potential determinant explaining soil temperature and moisture differences as a function of the mulch material used is the colour aspect. Dark-coloured surfaces, such as in *Leucaena leucocephala* mulch, will absorb more radiation compared to lighter coloured surfaces found in the other two mulch materials. As a consequence temperatures under a mulch layer of *Leucaena leucocephala* are relatively high; Table 2.

Both *Leucaena leucocephala* and *Gliricidia sepium* and have relatively low C/N ratios (c. 12) which favour fast decomposition by micro-organisms, compared to *Flemingia macrophylla* with a C/N ratio of 21 [3]. The latter leaf biomass was found to decompose about half as fast as the other two mulches during *in-vitro* digestibility trials, while the 'half-life' value of this mulch material under field conditions was about twice as long as the values found for *Gliricidia sepium* and *Leucaena leucocephala* [3].

In spite of this difference in decomposability of the materials, the decrease in the effect on soil temperature and moisture occurs at similar rates, since the values for the regression coefficients in the Figures 2 and 3 are about equal. Apparently chemical composition is not the only factor that determines the *effective life-time* of a mulch material with regard to soil moisture conservation and temperature reduction. In the case of the mulch of *Flemingia macrophylla*, a likely explanation of the similarity in the effect patterns found (see Figures 2 and 3) is that a relatively rapid collapse of structure and volume takes place, bringing the appearance of the decaying mulch layers closer to each other.

Of the three mulch materials tested that of *Flemingia macrophylla* clearly is the best-performing in terms of soil moisture conservation and temperature reduction. This result is consistent with the relative efficiency of the mulch materials in weed control [2].

6. Conclusion

The basic problem in this type of study is that results are site and situation-specific and a quantitative comparison of effects between sites cannot be justified: Soils differ widely in their capacity to contain and conduct moisture as well as heat, while prevailing climatic conditions (especially rainfall and temperature) during trials influence decomposition rates of organic materials. The linear decrease in the performance of the mulches found here should therefore not be taken as the general pattern of the influence of decomposing mulches on soil microclimatic parameters.

However, management of organic matter is a cornerstone in low-cost cropping system technology development for the humid tropics, warranting research on mulching as cultural technique and investigations into the behaviour of material of different origin. Results will gain in value describing trial conditions, e.g. rainfall, temperature and soil moisture content profiles, and when a certain mulch type of known composition (for example that of *Leucaena leucocephala*) is adopted as a standard to interpret the results of comparative mulching trials [3].

7. Acknowledgement

Thanks are due to Prof. C.J. Stigter for his valuable comments on earlier drafts of this paper.

References

1. Budelman A (1988) Leaf dry matter productivity of three selected perennial leguminous species in humid tropical Ivory Coast. Agroforestry Systems 7(1): 47–62
2. Budelman A (1988) The performance of the leaf mulches of Leucaena leucocephala, Flemingia macrophylla and Gliricidia sepium in weed control. Agroforestry Systems 6(2): 137–145
3. Budelman A (1988) The decomposition of the leaf mulches of Leucaena leucocephala, Gliricidia sepium and Flemingia macrophylla under humid tropical conditions. Agroforestry Systems 7(1): 33–45
4. Budelman A (in press) Nutrient composition of the leaf biomass of three selected woody leguminous species. Agroforestry Systems
5. Budelman A (in press) Effect of the application of the leaf mulch of Gliricidia sepium on early development, leaf nutrient contents and yields of Water yam (Dioscorea alata) Agroforestry Systems
6. Harrison-Murray RS and Lal R (1979) High soil temperature and response of maize to mulching in lowland humid tropics. In: Lal R and Greenland DJ Ed. (1979) Soil physical properties and crop production in the tropics. John Wiley & Sons. Chichester. p 285–304

7. Lal R (1975) Role of mulching techniques in tropical soil and water management. IITA Technical Bulletin no. 1. pp 37
8. Mellaart EAR (1975) Muls en microklimaat. De betekenis van een muls en de veranderingen van het temperatuursverloop tengevolge daarvan. Mededelingen van de Landbouw Hogeschool Wageningen 75–5 Agricultural University Wageningen, The Netherlands. pp 29
9. Othieno CO, Stigter CJ and Mwampaja AR (1985) On the use of Stigter's ratio in expressing thermal efficiency of grass mulches. Experimental Agriculture 21: 169–174
10. Sanchez PA and Salinas JG (1981) Low input technology for managing Oxisols and Ultisols in tropical America. Advances in Agronomy 34: 279–406
11. Siemonsma JS (1982) La culture du Gombo (Abelmoschus spp.) Légume-fruit tropical. Department of Tropical Crop Science, Agricultural University Wageningen, The Netherlands. pp 297
12. Snedecor GW and Cochran WG (1973) Statistical methods (6th ed.) Iowa State University Press. pp 593
13. Stigter CJ (1984) Mulching as a traditional method of microclimate management. Archives for Meteorology, Geophysics, and Bioclimatology, Serie B 35: 147–154
14. Stigter CJ (1985) Physics of mulching, with particular emphasis on grass mulches. In: Proceedings of the International Colloquium on Energy Flux at the Soil/Atmosphere interface, Trieste, Italy. ICTP, Trieste
15. Wade MK and Sanchez PA (1983) Mulching and green manure applications for continuous crop production in the Amazon Basin. Agronomy Journal 75: 39–45

Agroforestry Systems 8: 243–256, 1989

7
Effect of the application of the leaf mulch of *Gliricidia sepium* on early development, leaf nutrient contents and tuber yields of water yam (*Dioscorea alata*)

ARNOUD BUDELMAN
Department of Tropical Crop Science, Agricultural University, Wageningen, The Netherlands
Present address: Royal Tropical Institute, Mauritskade 63, Amsterdam, The Netherlands

Key words: Low External Input Agricultural Technology, *Gliricidia sepium*, leaf mulch, *Dioscorea alata*, yam sett germination, yam leaf nutrient content, yam tuber yields, humid Ivory Coast

Abstract. This paper describes the effects of the leaf mulch of *Gliricidia sepium* on the development and yield of Water Yam, *Dioscorea alata*. Using leaf mulch, the time taken for the yam setts to sprout can be shortened by approximately 20 percent.

Organic mulches contain considerable quantities of plant nutrients. Increasing amounts of mulch improved the leaf nutrient contents of the yam crop and resulted in significantly higher tuber yields. Over a tuber yield range up to c. 15 tons ha^{-1} each additional ton DM *Gliricidia sepium* mulch applied resulted in a yield increment of about 2 ton yam tubers. A nutrient supply – nutrient extraction balance is discussed, comparing mulch applied and yam tubers harvested.

Mulching as agricultural technique is a useful and affordable tool in adapting low external input cropping systems to local economic and environmental conditions.

1. Introduction

This paper is part of a series in which the use of three nitrogen-fixing perennials (*Leucaena leucocephala, Gliricidia sepium and Flemingia macrophylla*) is discussed in relation to the cultivation of Water Yam, *Dioscorea alata*. The overall aim of the research programme was to contribute to the development of a low-cost, sustainable cropping system including yam as foodcrop, part of a potential substitute for the practice of shifting cultivation.

Conceptually, the study falls within the context of the *Low External Input Agricultural Technology* development [21]. In this type of agricultural technology development solutions are sought to existing environmental constraints to agricultural production, such as soil acidity and low fertility, while making use of local resources including knowledge and experience of farmers.

A classical example of a low external input cultivation technology is the 'alley-cropping' system pioneered by the International Institute of Tropical Agriculture, in which *Leucaena leucocephala* features [11,12,18]. While the studies referred to are mainly concerned with the use of leaf mulches as a source of nitrogen, in a cropping system shrubs and trees potentially play a more diverse role in optimising conditions for crop growth. Making use of the various specific qualities of perennial species new or improved cropping systems can be designed, and appropriate management procedures can be devised. Such an excercise requires intimate knowledge of properties and aspects of the woody species involved; biomass productivity, nutrient content, quality of the leaf mulch [2,3,4], but also tree architecture, root area index and distribution, wood quality, *etcetera*[1].

In this paper the effect of the leaf mulch of *Gliricidia sepium* is discussed in relation to the life cycle of Water Yam (*Dioscorea alata*); establishment, mid-term nutrient status of the crop and tuber yields.

2. Location

The trial was conducted on the experimental farm on the ORSTOM research station near Abidjan, in the humid coastal zone of Ivory Coast. The farm is located on a well-drained sandy soil (clay content < 15%; pH-H_2O 4.7; CEC 3-5 Meq.; strongly desaturated, typically < 15%). The annual rainfall is about 2,000 mm. For more information on climate and origin of the woody perennials we refer to a previous publication on the productivity of leaf biomass of the species [2].

The field used for the experiment had been under yam cultivation for the third consecutive year.

3. Materials

Table 1 summarises data on the leaf mulch of *Gliricidia sepium* used in the experiments. The data on the nutrient content of the leaves of the woody perennial are averages over a period of observation of three and a half years [4].

In the lower row of Table 1 the values for the 'half-life' time of the nutrients are given, the time in days it takes to lose half of the nutrient load from a mulch layer [3].

[1] In preparation: 'Woody legumes as live support systems in yam cultivation'

Table 1. The nutrient content of the mulch and corresponding 'half-life' time during decomposition

Element	N	P	K	Ca	Mg	
Content	3.18	0.24	1.68	1.34	0.48	% of DM
Half-life time	21.7	19.8	11.0	28.9	16.5	days

[Source; ref. 3,4]

As Table 1 shows potassium is easily leached from the material, while calcium tends to stay behind. The relative rate of release of the nutrients is important in relation to the manurial capacity of the mulch material.

The yam species used as testcrop is *Dioscorea alata*, cv. 'Brazo fuerte'. This cultivar is known as being highly resistant against anthracnose, one of the dominant yam diseases in the area.

4. Trial design and sampling procedure (General aspects)

As lay-out of the field-trial a standard, completely randomised block design was chosen, including four replications. The experimental units measured 7 by 6 m. The net yam plant number per experimental units was 65. Yam setts are planted in rows, spaced 1 m apart, while the in-row distance between yam plants was 0.5 m.

The values for the *Least Significant Difference* are given at a confidence level of 95 percent. Where the average, overall effect of treatment is tested against the control data *Student's t*-test is applied; Table 4. Where average effects from treatments are compared among each other the *Studentised Range*-test is used; Tables 2,4,6. The latter test is identical to TUKEY's *Multiple Range*-test [22: p. 271].

Mature yam leaves were sampled to determine the crop nutrient status. Mature yam leaves included those leaves that were dark green in colour, but showed no yellowing, the visible sign of the onset of senescence. Six randomly chosen, complete yam tubers were analysed on the nutrient status. Leaf and tuber samples were analysed at the central laboratory of ORSTOM at Adiopodoumé, Ivory Coast [4]

5. The effect of mulching on early development of the yam crop

5.1 Method

Leaf mulch of *Gliricidia sepium* was applied just after the yam setts were planted, at rates of 200 –, 300 – and 600 g fresh material per sett. The mulch

Table 2. The effect of mulch application on sprouting and shoot development of yam setts (39 plants observed per experimental unit; 4 replications)

Treatment fresh mulch in g/plant	4 weeks		5 weeks		6 weeks	
	Nr. of plants emerged	Sprout length cm	Nr. of plants emerged	Sprout length cm	Nr. of plants emerged	Sprout length cm
Untreated	7.7	8	18.0	11	27.0	28
200*	18.3	14	26.8	28	32.3	57
300	19.5	17	29.0	32	32.5	61
600	21.5	25	31.7	42	31.2	75
LSD** (P = 0.05)	10.9	11.4	9.8	18.7	n.s.	23.4

* instead of 4 this treatment included 8 experimental units
** LSD according to the Studentised Range-test (Tukey)

material was concentrated into 50 cm wide strips over the yam rows. The application corresponds to 0.9, 1.3 and 2.6 ton ha^{-1} of leaf dry matter. Only 50% of the surface of the experimental units was actually covered by the mulch. The 'apparent" application rate therefore was 1.8, 2.6 and 5.2 ton ha^{-1}.

Of a total of 5 rows of yam plants available per experimental unit the 1st, 3rd and 5th row were measured each round. Emergence was noted and the length of the sprout measured.

5.2 Results

In Table 2, yam emergence and average sprout length are presented at the 4th, 5th and 6th week after planting the yam setts.

When mulch is applied about 50 percent of the yam setts are sprouted after 4 weeks. For the un-mulched plots this criterion is met about one week later; Table 2. Mulching as a cultural practice shortens the period of the yam sprouts to emerge by approximately 20 percent. The advantage of early sprouting in the mulched over the un-mulched experimental units is lost after a few weeks when sprouting is also completed in the latter situation. The advantage continues however to be visible when taking into account the lead in development of the yam plants grown under mulched conditions. Sprout length still differs significantly after 6 weeks.

5.3 Discussion

One of the characteristics of the yam culture is the slow establishment of a closed cover. Particularly during the first month of the cultivation cycle, soil cover is incomplete and soil erosion is likely to occur. In traditional cropping systems such hazard is partly avoided by planting mixtures of crops. Fast growing crop species, such as *Abelmoschus esculentus*, are sown early and provide soil cover before yam plants emerge and start claiming space[2].

Mulching as a cultural practice will enhance the quality of soil cover early in the cropping cycle. Its effect is twofold. Firstly, there is a *direct* influence of the presence of a layer of material which limits the destructive impact of raindrops on the soil structure and increases the capacity of the soil to absorb water.

There is as well an *indirect* effect of a mulch layer. Table 2 shows that mulching promotes early development of the yam plants. Within the range of treatments the effect of the application of the leaf mulch of *Gliricidia sepium* is significantly different from the untreated experimental plots at the level of 300 g per plant fresh material. Doubling that quantity does not further improve the sprouting rate of the yam plants. In other words, also moderate amounts of mulch material make sense; 1.3 ton ha^{-1} DM for 20,000 plants.

The positive influence of mulches on the emergence of yam has been reported before [14,15,17]. Three possible reasons may explain the better initial development of the yam crop when mulch is applied. Nutrients are released from the decomposing mulch which may enhance early growth, although the initial development is known to be based largely on the nutrients present in the yam setts [17].

A more likely explanation is found in the influence a mulch layer exerts on the moisture content and temperature in the soil. Mulched soils tend to have a higher content, particularly important during dry spells.

Table 3 illustrates the influence of a *Gliricidia sepium* mulch layer on the temperature in the soil and the degree to which evaporation losses can be avoided.

The sandy soil on which the experiment took place has a low water storage capacity; from 3.7% ('wilting point') to 14.2% ('field capacity') [5]. The low upper limit is not uncommon as Salter & Williams (1965) have pointed out in a general review on the relations between soil texture and water storage capacity.

Particularly on sandy soils mulching is an important means to improve the micro-climate in the cropping environment. As Table 3 shows the mulch layer of *Gliricidia sepium* provides an average of 3.9% extra moisture over the first 5 cm of the soil compared to the situation in a bare soil. This additional moisture counts for 37% of the range of available moisture.

2 In preparation: 'Land-use by the immigrant Baoulé tribe in the Taï region, South-west Ivory Coast'

Table 3. Reduction of evaporation losses and soil temperature as a function of the presence of a *Gliricidia sepium* leaf mulch layer at 5 ton per hectare dry matter

Effect	Average over 60 days	Initial impact at t = 10 days	Estimated duration of the effect
Temperature at 5 cm depth*	−4.6 °C	−7.5°C	65 days
Additional soil moisture 0–5 cm*	+3.9%	+5.1%	80 days

*relative to the situation in a bare soil. [Source: ref. 5]

In spite of the fact that *Gliricidia sepium* mulch decomposes relatively fast – half of the dry matter is lost within three weeks of application [3] – the moisture conserving effect lasts long enough to cover the period during which the crop completes its establishment.

A second important effect of mulches is their capacity to reduce the temperature in the upper soil strata. Table 3 shows that the average maximum soil temperature at 5 cm depth is 4.6 °C lower than in the case of an un-mulched soil: 32.5 °C against 37.1 °C. The initial effect is even larger (7.5 °C) and if early yam sprouting is negatively affected by high ambient temperatures mulching is a simple cultural practice to avoid damage. Although no information has been found concerning the optimum tem-

Table 4. Average nutrient content in the yam leaves as a function of the amounts of *Gliricidia sepium* mulch applied (4 replications: mature yam leaves sampled)

Leaf DM ton per hectare	Nutrients				
	N %	P %	K %	Ca %	Mg %
Untreated	2.29	0.37	1.23	0.85	0.32
1.6	3.11	0.33	1.65	1.17	0.38
2.8	2.35	0.37	2.16	1.12	0.40
3.8	3.74	0.36	2.47	1.18	0.40
5.7	3.00	0.35	2.99	0.95	0.42
Tukey's LSD (P = 0.05)*	0.87	NS	0.93	0.22	0.09
Untreated versus	2.29			0.85	0.32
average all treatments	3.05			1.10	0.40
Student's LSD (P = 0.05)**	0.47			0.12	0.05

*testing individual treatment differences (4 samples of treatment 1 vs. 4 samples of treatment 2, etcetera)
**testing 'untreated' (4 samples) against 'treated' (16 samples)

perature range for the early stages of development of yam plants, maize seedlings e.g., grown under controlled conditions, are known to virtually cease growing when soil temperatures rise to over 36°C [13].

It is difficult, based upon the observations done and information available, to draw conclusions on the relative importance of the two factors discussed with regard to the early development of the yam sprouts. In the field the effect of the two factors, temperature and soil moisture, is strongly confounded and not easily studied as separate components [8].

6. The effect of mulching on the nutrient content in yam leaves

6.1 Method

After the initial mulch application at planting time (see par. 5.1) the experimental units were regularly provided with known quantities of leaf mulch lopped from a nearby stand of *Gliricidia sepium.*

The yam crop took four months to arrive at a closed cover in the plots that received mulch treatments. The total mulch quantities applied were respectively 1.6, 2.8, 3.8 and 5.7 ton ha^{-1} leaf dry matter of *Gliricidia sepium.*

The general nutrient status of the yam crop at mid-term (5.5 months after planting) was determined through sampling and nutrient content analysis of mature yam leaves.

6.2 Results

Table 4 shows the concentrations of the major nutrients in mature yam leaves as a function of increasing amounts of *Gliricidia sepium* leaf mulch applied. With exception of phosphorus a positive effect was noted ($P < 99.5\%$). The most consistent response to increasing quantities of mulch was found in the case of potassium. The effect of different amounts

Table 5. Estimated percentage of nutrients released from a mulch layer as a function of time (figures for 4 ton ha^{-1} leaf DM from *Gliricidia sepium*)

Time	Nutrient	N	P	K	Ca	Mg
1 week		20%	22%	36%	16%	26%
2 weeks		36	39	59	29	45
4 weeks		59	63	83	49	69
6 weeks		74	77	93	64	83

[source: ref. 3; equation 2]

of mulch applied on the concentrations in the leaves of nitrogen and calcium, although partly statistically significant, was erratic.

6.3 Discussion

Nutrients from mulches are generally slow to move from the original matrix compared to nutrients in easily soluble inorganic fertilisers. On the one hand this may be an advantage, for example during heavy showers. Mulches then offer some protection against strong leaching and transport of nutrients through surface run-off. On the other hand the slow release of nutrients from the mulch matrix may result in low efficiencies in terms of providing a crop with essential nutrients. This has been observed with *Leucaena leucocephala* prunings as a source of nitrogen in maize production [11,12,18].

Table 5 shows the nutrient release pattern from the mulch of *Gliricidia sepium*, based on a nutrient release function developed from data collected in a decomposition study which took place during the rainy season [3].

Knowing dates and quantities of the mulch dressings Table 5 offers the possibility to estimate the amounts retained in the mulch residue. At the moment of yam leaf sampling the estimated amount of nutrient left in the *total* quantity of mulch applied over the first 4 months of the cultivation cycle varied from 2% (potassium) to 13% (calcium). Except for calcium more than 90% of the nutrient load in the mulch layers was released.

The results presented in Table 4 show three response patterns;

i)Mulching does not lead to increased leaf nutrient content. This is the case for phosphorus, with an average leaf nutrient content of 0.36 percent. This level is probably sufficient. Obigbesan & Agboola [1978] found for the leaves of *Dioscorea alata* an average phosphorus content of 0.35%. The leaves sampled came from a trial that received a basal dressing of 90 kg ha^{-1} of P_2O_5.

ii)In the second response pattern, mulching results in an increase in the nutrient content in the yam leaves, but responses are erratic. This pattern is found in the cases of nitrogen, calcium and magnesium. For these nutrients untreated leaf samples are tested against a general average from the treated samples; Table 4.

iii)The only nutrient of the group studied that shows a significant linear relationship between leaf nutrient content and quantity of mulch applied is potassium; Figure 1.

The response of the yam crop to mulching clearly reflects the soil fertility profile of the soils of the experimental area as described by Roose &

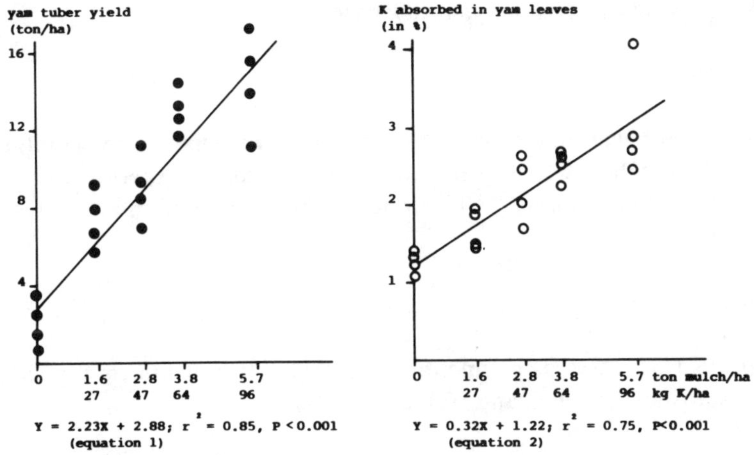

Fig. 1. The effect of mulch application on yam tuber yield (fresh weight) and on potassium content in the yam leaves.

Cheroux [1966]. Normally, the soils are sufficiently well-stocked in terms of nitrogen and especially phosphorus. In the case of nitrogen the erratic response by the yam crop to mulch may stem from the fact that during decomposition of the mulch nitrogen losses occur through volatilisation.

Potassium is present in limited to occasionally very limited quantities; 0.01–0.05 Meq., while calcium and magnesium are found in the range of 0.2 to 0.6 Meq. per 100 g soil [19].

Response patterns and information on nutrient availability in the soil point at potassium as the most limiting nutrient for yam production, a conclusion supported by the results from experiments conducted on leached soils of the forest region in Eastern Nigeria [9].

This conclusion however may have a limited validity. In a recent review Degras [1986] discusses fertiliser studies in which yam failed to produce a response to the application of potassic fertilisers. This was attributed to differences in the origin of parent material of soils, and the fact that yam species and even cultivars differ in their requirement of potassium.

Table 6. The effect of increasing quantities of dry matter leaf mulch of *Gliricidia sepium* on the yam tuber yield

Treatment	Control	1.6	2.8	3.8	5.7	ton ha^{-1}
average tuber yield;	1.96	7.21	8.87	12.87	14.49	ton ha^{-1}
Statistical significance: $P < 0.001$; LSD ($P = 0.05$)* = 2.94 ton ha^{-1}						

*LSD-value according to the *Studentised Range*-test

7. The effect of mulching on yam tuber yields

7.1 Method

At the end of the cropping cycle the yam tubers were harvested. Total fresh weight of the tubers was determined for each of the 20 experimental units, and converted into ton ha^{-1}. The latter figures were used in an analysis of variance.

7.2 Results

Tuber yield increases as a function of the quantity of mulch applied are highly statistically significant; Table 6.

7.3 Discussion

The discussion in paragraph 6.3 ended with the conclusion that potassium can be considered as the most limiting nutrient in case of our experiment. Under given conditions of soil fertility yam tuber yields and the leaf potassium content showed a strong positive correlation ($r = +0.82$). The element therefore appears to be an appropriate indicator to determine the manurial value of the mulch of *Gliricidia sepium* in yam cultivation on soils low in available potassium.

Table 7. A nutrient balance

	Nutrients in kg *extracted* via yam harvest; based on 0.533 ton DM yam tubers)					
	N	P	K	Ca	Mg	
Ivory Coast[a)]	5.44	0.75	5.12	0.27	0.43	kg
Nigeria[b)]	7.57	1.00	9.54	0.16	0.47	
Trinidad[c)]	7.30	0.70	6.18	0.50	0.43	
	nutrients in kg *supplied* by *Gliricidia* mulch; based on 1 ton DM leaf mulch)					
Ivory coast[d)]	31.8	2.4	16.8	13.4	4.8	kg
other sources[e)]	38.2	2.3	28.0	14.2	4.6	
supply (:) extraction ratio (this study)	*5.8*	*3.2*	*3.3*	*49.6*	*11.2*	

[a)] tuber nutrient content N 1.02%, P 0. 14%, K 0.96%, Ca 0.05%, Mg 0.08% [b)] data derived from Obigbesan & Agboola [1978]; [c)] data derived from Ferguson et al. [1980]; [d)] ref. 4; [e)] compiled in ref. 4

In Figure 1 tuber yields and potassium content of the leaves are presented as a function of the quantity of mulch applied. Regression equation and correlation coefficient concerning tuber yields show that of the variation in the tuber yields (Y) 85% can be attributed to its linear regression on X, the quantity of mulch applied.

From regression equation 1 in Figure 1 it can be deduced that each ton of dry matter leaf mulch of *Gliricidia sepium* results in a tuber yield increment of 2.23 ton. This relation is valid for a tuber yield range up to c. 15 ton ha^{-1}. The tuber yield increase in eq. 1 is based on the fresh weight of the tubers.

From randomly selected complete yam tubers the dry weight (:) fresh weight conversion factor was established at 0.239. The yield increment of 2.23 ton fresh yam tubers then represents 0.533 ton dry matter.

In Table 7 a comparison is made between the amount of nutrient supplied through mulch and the amount extracted when the yam harvest is transported from a field. The information from this study is supplemented by data from comparable research.

The yam tubers in this study contain less nitrogen (1.02% against an average of 1.39%) and potassium (0.96% against 1.47%) compared to the two other studies. This agrees with the situation at the supply side. The average nitrogen content in the leaves of the mulch of *Gliricidia sepium* in the compilation from literature was considerably higher than found in the leaf mulch raised in our area of study: 3.82% against 3.18 percent. A similar conclusion can be drawn concerning the data on potassium in the leaves of *Gliricidia sepium:* 2.80% as average K-content from the literature survey against a mere 1.68% in the leaves of the trees in our study [4].

The differences in the potassium content in the leaves of both the yam crop and mulch-producing tree species certainly reflect the paucity of this particular element in the soils where the trials were conducted [19].

The supply (:) extraction ratios in Table 7 show that the mulch of *Gliricidia sepium* supplies sufficiently large amounts of the 5 macro-nutrients to compensate for the losses through harvest and transport of the yam tubers. This is true when the field where the mulch is raised is separated from the place where it is used, the yam field. A field of *Gliricidia sepium* trees, however does not 'produce' nutrients, with the possible exception of some of the nitrogen that is fixed through bacterial activity in the root nodules [4]. Consequently, at the level of the farming system, one field is mined with the help of the trees, maintaining the soil fertility status elsewhere.

When the trees are grown on the same field in association with the yam crop, the latter may profit from nutrients mobilised by the trees when these are absorbed from soil layers not normally exploited by the yam roots (see reference footnote 1).

On the long run in both scenarios for the use of trees depletion of soil nutrients may take place, if losses exceed the natural influx (e.g. from dust) and no compensation is sought for the losses caused by the removal of the tuber yield.

8. Conclusions

The experiment was conducted on a sandy soil, particularly poor in available potassium. Such type of soil is generally considered ill-suited to yam production. Moreover the same field had been cultivated for three successive years, including this experiment. The response to mulching in terms of quantity of tubers produced (one ton of *Gliricidia sepium* leaf DM results in 2.23 ton tubers; Table 7) therefore must be considered as approaching the maximum possible.

In a trial discussed elsewhere (see reference footnote 1), on a field previously under a grass fallow, 4.8 ton ha^{-1} DM leaf mulch produced an additional 9.0 ton yam tubers, or c. 1.9 ton tubers per ton *Gliricidia sepium* mulch. The control in that experiment yielded 8.8 ton ha^{-1} yam tubers.

In the context of yam cultivation on soils with a low chemical fertility a response rate of c. 2 ton of tubers per ton DM leaf mulch of *Gliricidia sepium* is proposed as a base for calculation of the economy of mulching as a crop husbandry technique.

The trial results show that the use of *Gliricidia sepium* as a *single* cultural practice has at least three important effects in a cropping environment: i) lower maximum temperature in the upper soil strata, ii) higher soil moisture content, and iii) an improved nutrient supply. A fourth effect, the influence of mulch application on weed growth, is discussed elsewhere [1].

All aspects add to a situation closer to an optimum for crop growth as compared to the condition in an un-mulched soil. Because of these accrued benefits a substantial improvement in yam tuber yields is possible.

Under field conditions the relative contribution of each of the aspects mentioned to the final yield achieved is difficult to evaluate separately because of interdependency [8]. Nutrient supply – e.g. nitrification during decomposition of organic mulches – is intricately bound to both the ambient soil moisture as well as temperature. From that perspective it seems appropriate to let the problems specific in the cultivation of yam as such dictate the moment of mulch application.

Especially early in the yam cultivation cycle mulches are valuable as soil cover and help to enhance the acceptance of rainwater. This protective action is often considered as being the principal benefit from mulching as an agricultural practice [10]. In that choice the maximum benefit in terms of water conservation and lower soil temperatures is automatically assured, while the crop can profit at a later stage from the nutrients released.

As a general conclusion, mulching as practice combines several advantages that are worthwhile to consider when searching to adapt cropping systems to local ecological and socio-economic conditions: i) mulches play a multi-functional role in crop environmental management; ii) the technique of mulching is inherently simple and therefore user-friendly; iii) mulching is within reach of resource-poor farmers as long as land and labour are available to produce material. Economically speaking, the practice has a low treshold compared e.g. to fertilisers.

References

1. Budelman A (1988) The performance of the leaf mulches of Leucaena leucocephala, Flemingia macrophylla and Gliricidia sepium in weed control. Agroforestry Systems 6: 137–145.
2. Budelman A (1988) Leaf dry matter productivity of three selected perennial leguminous species in humid tropical Ivory Coast. Agroforestry Systems 7: 47–62.
3. Budelman A (1988) The decomposition of the leaf mulches of Leucaena leucocephala. Gliricidia sepium and Flemingia macrophylla under humid tropical conditions. Agroforestry Systems 7: 33–45.
4. Budelman A (in press) Nutrient composition of the leaf biomass of three selected woody leguminous species. Agroforestry Systems.
5. Budelman A (in press) The performance of selected leaf mulches in soil temperature reduction and soil humidity conservation. Agroforestry Systems.
6. Degras L (1986) L'igname. Plante à tubercule tropicale. Editions GP Maisonneuve & Larose. pp 408.
7. Ferguson TU, Haynes PH and Spence JA (1980) Distribution of dry matter and mineral nutrients in tubers of two cultivars of Dioscorea alata L. Tropical Agriculture (Trinidad) 57 (1): 61–67.
8. Harrison-Murray RS and Lal R (1979) High soil temperature and the response of maize to mulching in the lowland humid tropics. In: Lal R and Greenland DJ (Ed.) Soil physical properties and crop production in the tropics. John Wiley & Sons, Chichester, UK. p. 285–304.
9. Irving H (1956) Fertilizer experiments in Eastern Nigeria: 1947–1951. Tropical Agriculture (Trinidad) 33 (1): 67–78.
10. Jones PA, Robinson JBD and Wallis JAN (1960) Fertilizers, manure, and mulch in Kenya coffee growing. Empire Journal of Experimental Agriculture 28 (112): 336–352.
11. Kang BT, Sipkens L, Wilson GF, and Nangju D (1982) Leucaena (Leucaena leucocephala (Lam.) de Wit) prunings as nitrogen source for maize (Zea mays L.) Fertilizer Research (The Netherlands) 2 (4): 279–287.
12. Kang, BT, Wilson GF and Sipkens L (1981) Alley cropping maize (Zea mays L.) and

Leucaena (Leucaena leucocephala Lam.) in Southern Nigeria. Plant & Soil (The Netherlands) 63 (2): 165–179.
13. Lal R (1974) Effects of constant and fluctuating soil temperature on growth, development and nutrient uptake of maize seedlings. Plant and Soil (The Netherlands) 40: 589–606.
14. Lal R and Hahn SK (1973) Effect of method of seedbed preparation, mulching and time of planting on yam in Western Nigeria. Paper presented at the 3rd International Symposium on Tropical Root Crops, Ibadan, Nigeria. p. 293–306.
15. Maduakor HO, Lal R and Opara-nadi OA (1984) Effects of methods of seedbed preparation and mulching on the growth and yield of White Yam (Discorea rotundata) on an Ultisol in South-East Nigeria. Field Crops Research 9: 119–130.
16. Obigbesan GO and Agboola AA (1978) Uptake and distribution of nutrients by yam. Experimental Agriculture 14: 349–355.
17. Onwueme IC (1978) The tropical tuber crops, Yams, Cassava, Sweet Potato and Cocoyams. John Wiley & Sons, New York. p 3–106.
18. Read MD, Kang BT and Wilson GF (1985) Use of Leucaena leucocephala (Lam.) de Wit leaves as a nitrogen source for crop production. Fertilizer Research (The Netherlands) 8 (2): 107–117.
19. Roose E and Chéroux M (1966) Les sols du bassin sédimentaire de Côtc d'Ivoire. Cahier ORSTOM-Série Pédologie 4 (2): 51–92.
20. Salter PJ and Williams JB (1965) The influence of texture on the moisture characteristics of soils. II. Available water capacity and moisture release characteristics. Journal of Soil Science 16 (2): 310–317.
21. Sanchez PA and Salinas JG (1981) Low input technology for managing Oxisols and Ultisols in tropical America. Advances in Agronomy 34: 279–406.
22. Snedecor GW and Cochran WG (1973) Statistical methods (6th ed.) Iowa State University Press. pp 593.

IV Integrating woody perennials in annual cropping systems

Agroforestry Systems 10: 47–59, 1990.

8
Woody legumes as live support systems in yam cultivation
1. The tree-crop interface

ARNOUD BUDELMAN
Centre Néerlandais, Abidjan. Côte d'Ivoire. Department of Tropical Crop Science, Agricultural University Wageningen, The Netherlands
Present address: Royal Tropical Institute, Mauritskade 63, 1092 AD Amsterdam, The Netherlands

Key words: *Gliricidia sepium, Leucaena leucocephala, Flemingia macrophylla,* live support systems, Water Yam (*Dioscorea alata,* cv. Brazo Fuerte), plant architecture, root systems

Abstract. This paper reports on the use of three leguminous perennials (*Leucaena leucocephala, Flemingia macrophylla* and *Gliricidia sepium*) as live support systems in yam cultivation. In a planting arrangement in which yam rows alternated with rows of the woody species the tuber yields per ha were 3.4 (leucaena), 5.3 (flemingia) and 10.1 (gliricidia) ton fresh weight. TUKEY's L.S.D. value equalled 2.9 ton.

Leucaena leucocephala is unsuitable as live support since the species shows a strong competitive power expressed in terms of leaf productivity and relative density of the root mass in the upper soil *stratum*, the zone explored by the yam crop.

Flemingia macrophylla is unsuitable as support species mainly because of its structural weakness: Its branches do not sufficiently lignify to carry the yam leaf mass.

The significantly higher tuber yield of the yam crop grown with *Gliricidia sepium* is a function of specific properties of the tree species: Low leaf productivity, a relatively weakly developed root system and an open architecture, which leaves sufficient space for a yam crop grown in association.

1. Introduction

Yams are a highly appreciated staple in West Africa. In recent years production has, however, considerably decreased. One of the reasons is that the crop requires staking to achieve economically viable production levels. In many areas, as a consequence of deforestation, stakes are difficult to obtain, or are prohibitively expensive when transported from other regions.

As a result yam is no longer able to compete on the market, and is rapidly losing its share to other tuber crops that can be raised cheaper, such as cassava.

One solution would be to replace the traditional cultivars by cultivars that produce well without staking, and to mechanise its cultivation [15]. This must be taken as a long term perspective.

On the short term intermediate solutions could be studied and proposed to farmers. These include the production of stakes on wastelands, as well as the improvement of live support systems, the integration of suitable tree species in the yam cropping system itself [16].

This paper reports on the use of three woody leguminous perennials as possible candidates in live support systems; *Gliricidia sepium, Leucaena leucocephala* and *Flemingia macrophylla*. The first two species are trees, the latter species is a shrub. The species all belong to the plant Family of the *Leguminosae*.

Gliricidia sepium was introduced during colonial times in Nigeria as shade tree in cacao plantations, and seems to be naturalised in the mean time. The tree, which roots readily from cuttings is used by farmers as live support in yam cultivation [1]. That practice may have been the origin of the significant role the tree plays nowadays in the woody fallow vegetation in South-west Nigeria [9], where the tree seems to be able to outcompete encroaching *Imperata cylindrica* [2,9].

In West Africa the other two species have not met farmers' acceptance. Both, however, have promising aspects, especially in terms of biomass productivity and nitrogen fixation [4,13]. Furthermore, the leaves of *Flemingia macrophylla* have been found to decompose relatively slowly, and its mulch can be successfully applied in weedcontrol [3], soil water conservation and soil temperature reduction [7].

The aim of this study is to compare the degree to which the three leguminous perennials are compatible with the growth habits and requirements of the yam plant. The focal area is the place where tree and crop interact. Is the combined effect of the interaction positive, planting must be arranged in ways that optimise interaction; *associated* cropping. Is, on the contrary, the overall effect negative, interaction should be limited, following *zonal* planting arrangements, e.g. 'alley-cropping', whereby the alleys are wide enough to avoid unwanted interference.

Field observations on yield and general plant characteristics can help to determine the planting arrangements distinguished. Such information is basic to the design of cropping systems including trees and annual crops [11].

2. Location and materials

2.1 Site of the trials

Experiments in which yams and the shrub/tree species were cultivated together started early 1984 on an experimental farm near Abidjan, Ivory

Coast. The farm is located on a well-drained sandy soil (clay content < 15%; pH 5; CEC 3–5 Meq. per 100 g soil; base saturation < 15%).

The annual rainfall is about 2,000 mm, distributed over one long rainy season (May to July) and a short rainy season (October to November). More information on climate and soil can be found in a previous publication [4].

2.2 Plant species used

Seeds from *Leucaena leucocephala* were collected in roadside vegetation near the experimental farm, where the species is found as a coloniser of eroded soils. The growth form of the species resembles the 'Hawaiian' type that is found as a common tropical weed outside its area of origin.

Material from *Gliricidia sepium* (cuttings) and *Flemingia macrophylla* (seeds) was acquired from the Coffee and Cacao Research station (IRCC) at Divo, where both species are used to provide shade and mulch in experimental coffee plantations.

For data on leaf productivity, leaf nutrient composition and leaf decomposition for the three perennial species see references 4,5 and 6 respectively.

As testcrop yam (*Dioscorea alata,* cv. Brazo Fuerte) was used. The cultivar was recommended by the *Institut des Savanes* (IDESSA) at Bouaké, from where planting material was obtained.

3. Methods

Data brought together in this paper have various origins (experiments, observations, results reported previously);

i) a hedgerow experiment in which yam was cultivated in association with the perennial species. This trial was conducted only once,
ii) light measurements in the crowns of *Gliricidia sepium* and *Leucaena leucocephala,*
iii) a study of the distribution of the roots of the three perennial species,
iv) qualitative observations on the form and extent of the root system of yam,
v) leaf productivity measurements concerning the three woody legumes.

3.1 The hedgerow experiment

Data on the yam tuber yields as a function of the hedgerow species used come from an experiment that was laid out as a completely randomised block design with 4 replications. In this experiment yam rows alternated with rows of the woody legumes. The rows stood 1 m apart. The 2 m distance between the hedgerows should not be taken as a recommendation. The choice, substituting one yam-row by a row of shrubs, allows 'two-way' exposure of the yam plants to possibilities to climb, hence to enlarge the leafmass.

The hedgerows were 7.2 m long, with an in-row spacing between the shrubs of 1.2 m. The planting arrangement corresponds with a planting density of c. 4,120 trees or shrubs ha^{-1}. Per experimental unit 4 rows of perennials were planted. At the onset of the experiment the hedges were 2 years old.

Each of the three alleys contained 13 yam plants, or 39 plants per experimental unit, border plants excluded. Yam setts were planted in flat soil at an in-row spacing of 0.5 m.

Mid-April the shrubs were stripped of all green matter. The biomass was weighed and applied as mulch. Yam setts were subsequently planted. Regrowth of the hedgerows was regularly checked in order not to hamper the developing yam plants. From the fourth month onward several branches of the shrubs were allowed to continue growing and to act as support for the yam vines.

The yam tubers were harvested mid-December and fresh weight of the tubers was determined.

3.2 Observations on light transmission in the tree crowns

Photosynthetically active radiation (PAR = 400 to 700 nm) in the crowns of *Gliricidia sepium* and *Leucaena leucocephala* was measured with a 'Licor' light meter, model LI—185 A provided with a quantum sensor, both made by the LAMBDA Instrument Corporation. Light was measured in the canopies along a straight line at fixed intervals of 10 cm, 150 cm above groundlevel, during a cloud- and windless day. Data variation has been limited using running means.

The measurements were repeated several times. One of the data sets is presented in order to demonstrate the nature of the light profiles.

3.3 The root systems of the woody species

Root systems of the three leguminous perennials were studied in three fan-shaped spacing trials that served to measure species leaf productivity [4]. At the moment of this study the stands were nearly four years old, and well-established.

For each of the species 7 plant densities were selected, which ranged from approximately 1,200 to 9,000 plants per ha. In each density class the root system of 5 shrubs was excavated and collected. The amount of soil surface sampled around each individual plant was a function of the plant density: The circular areas sampled had radii increasing (step 10 cm) from 30 (9,000 plants ha^{-1}) to 90 cm (1,200 plants ha^{-1}). This sampling procedure covered from a maximum of 30% (at 1,200 plants) to a minimum of 25% (at 9,000 plants ha^{-1}) of the total area available per plant in the trial plantation.

Excavation only took place after showers had moistened the soil sufficiently. The sandy soil (a minimum of 85% parts > 50 μm) permitted a high recovery of the roots. The soil around the plants was removed in layers of 10 cm, up to a depth of 40 cm. The roots were carefully separated from the soil with the help of a sieve with a 5 mm mesh. The rootmass collected was air-dried at 50° C during a week. After this period soil particles were gently removed from the root material. The roots were then divided into three categories; < 1 mm, between 1-5 mm and over 5 mm. Each category was weighed. Root samples were taken and oven-dried at 105° C, thus to provide the necessary corrections on the figures acquired at 50° C. The weight of the rootmass collected was converted into weight per m^2 soil surface.

3.4 The yam root system

The general form and extent of the root system of the yam was studied qualitatively just before harvesting in a field in which yams were planted at a 1 m by 1 m spacing.

3.5 Leaf productivity data of the hedgerow species

See reference 4.

4. Results and discussion

4. 1 Yam tuber yields as a function of the hedgerow species used

Table 1 shows the yam tuber yield as a function of the hedgerow species used. The tuber yield that results from the association with *Gliricidia sepium* as live support system differs significantly from the yield levels found in the other two combinations tested.

Since the mulch produced and distributed in the alley-cropping systems during the cultivation cycle was kept at about equal DM levels per treat-

Table 1. Yam tuber yields in three alley-cropping situations

Association	yam tuber yields in ton ha^{-1}
yam grown with *Leucaena leucocephala*	3.4^{a}
yam grown with *Flemingia macrophylla*	5.3^{a}
yam grown with *Gliricidia sepium*	10.1^{b}
Least Significant Differences (P = 0.05)	2.9

notes to Table 1:
-results with the same superscript do not differ significantly;
-L.S.D.-value based on TUKEY's criterion for significance.

ment, and trimming of the hedges was regularly and carefully executed, yam productivity differences largely stem from specific characteristics of the perennials that form the hedges: Plant architecture and structure, root systems, biomass productivity per unit of area.

4.2 Architecture and structure of the woody legumes

Several aspects that explain differences in suitability of the woody species as live support systems in yam cultivation have to be presented in more or less descriptive terms. Firstly, there is the general *habitus* of the species and a number of structural qualities related to this topic.

Among the species considered *Flemingia macrophylla* is the only true shrub, that is to say a woody perennial without a trunk. Although the management system enforced on *Gliricidia sepium* and *Leucaena leucocephala* caused a similar low bushy *habitus* with little or no trunk formation, both species retain the possibility to become a true tree, trunk and all.

The shrub *Flemingia macrophylla* consists of many upright branches arranged in a tussock. Although the branches easily can reach 2 m in height, little lignification occurs. This explains the tendency of the species to temporarily layer after heavy showers. When the yam shoots climb into the shrub its branches are permanently bent, and vegetative development of the yam in that case does not differ much from the situation in which support plants are absent.

Mature *Flemingia macrophylla* plants are densely foliated, forming a solid structure when arranged in hedgerows. Known as a generally sun-loving species [14], such circumstances are hardly inviting to the yam plant to explore. Consequently, limited use is made by the yam plant of the branches of *Flemingia macrophylla* for support.

A yam-*Flemingia macrophylla* alley-cropping system can be characterised by two separate compartments or zones, with little visible interaction. One

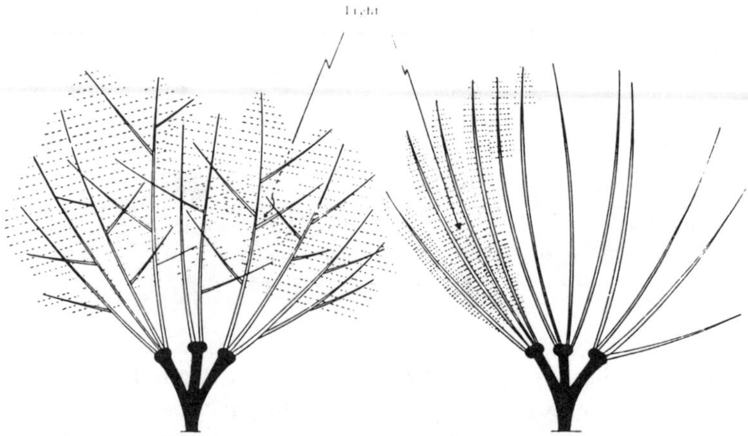

Fig. 1. Schematic presentation of the regrowth patterns in *Leucaena leucocephala* and *Gliricidia sepium*

zone is occupied by the yam plants, the other is almost exclusively used by the rows of *Flemingia macrophylla*.

In the cases of *Gliricidia sepium* and *Leucaena leucocephala* the use by the yam plants of the branches for support is quite intensive as compared to *Flemingia macrophylla*. Both species clearly have a less contracted, more open architecture, and are therefore inviting to the yam plant to enter. Budowski *et al.* (1984) citing various sources, mention both *Leucaena leucocephala* and *Gliricidia sepium* as species with canopies that permit considerable amounts of light to penetrate, but they note the absence of data to substantiate the observation. Our data suggest that the two species are far from equally successful in a role as a live support system. After being cut *Gliricidia sepium* produces a number of shoots, that initially continue growing virtually without branching; Figure 1. On the branches compound leaves are arranged in spirals, that eventually shed when the branches age.
The long, upright branches thus resemble arching, small-diameter columns. This typical structure is basically unstable since the centre of gravity is relatively highly place. Even gentle winds provoke considerable movement of the branches, and sunlight can penetrate deeply into the centre of the tree, where it creates changing patterns of sunflecks or 'dappled sunshine' [8]. Although the patterns constantly move, the potentially unfiltered nature of the light benefits yam vines that grow into the branches of *Gliricidia sepium.*

Leucaena leucocephala on the contrary branches regularly, and the compound leaves form partly overlapping layers. Consequently, the leaves in this species are probably more evenly distributed in the crown, effectively preventing incoming light from reaching the centre unfiltered; Figure 1.

This is further illustrated in Figure 2 where an example is given of the extinction of photosynthetically active light in the canopy of both species. The case presented comes from two regularly developed, free-standing trees with a biomass that equalled a 4-months old regrowth. The trees were about equally high. The difference in the light profiles which matters is that the canopy of the *Gliricidia sepium* tree shows gaps through which photosynthetically active light can reach the core of the tree, which potentially benefits a yam plant grown aside that uses the tree as support.

Part of the explanation regarding yam tuber yield differences is related to the wood quality of both species. After the fourth month, in the cases of *Gliricidia sepium* and *Leucaena leucocephala*, several branches per individual plant were left to grow and act as support for the yam vines. In this way climbing shoots of the yam plants could develop considerably better than when no support was available.

However, the wood of *Leucaena leucocephala* is rather solid, even in the early stages of regrowth. With *Leucaena leucocephala* support an increase of diameter of the branches causes yam vines to break, resulting in losses of leafmass of the yam in the upper reaches of the living tree branches.

The branches of *Gliricidia sepium* on the contrary stay green and soft, at least as long as it takes the yam crop to mature. Consequently, no broken

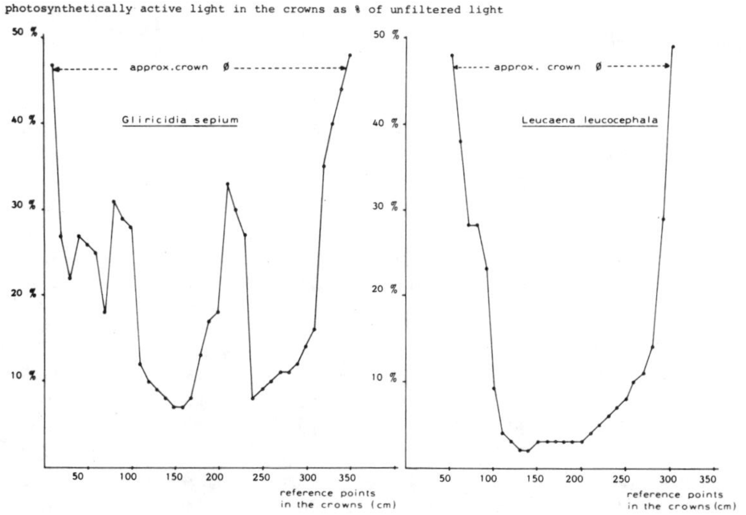

Fig. 2. Examples of light profiles in the crowns of *Gliricidia sepium* and *Leucaena leucocephala.*

vines were encountered in the stand where *Gliricidia sepium* was used as support.

This particular problem can be solved by ringing the bark of the *Leucaena leucocephala* branches at the basal side. This involves however a considerable labour input.

A disadvantage of the relatively soft wood of *Gliricidia sepium* is that occasional strong winds may cause the live poles to collapse, particularly at the end of the yam cropping cycle when the yam leafmass forms a considerable additional weight.

4.3 The root systems of the perennials

Of the data collected in the root system study only the fine root (< 1 mm)

Table 2. Average fine root (< 1 mm) dry weight in the upper soil strata (0–40 cm) for the three woody perennials (5 plants sampled per density, plants c. 4 years old)

		0–10	10–20	20–30	30–40 cm	total (g.m^{-2})
Gliricidia sepium;						
plants ha^{-1}	1080	19.0	15.7	8.0	3.2	45.9
	1410	22.4	20.6	8.4	2.7	54.1
	1880	24.7	19.4	11.5	5.2	60.8
	2560	25.0	16.1	15.0	9.4	65.5
	3620	24.0	18.6	12.2	8.1	62.9
	5360	23.5	21.2	10.7	8.2	63.6
	8260	30.1	13.8	7.8	9.6	61.3
average;	3450	24.1	17.9	10.5	6.6	59.1
Flemingia macrophylla;						
plants ha^{-1}	1180	13.8	8.2	3.1	1.2	26.3
	1535	15.1	9.4	5.8	1.3	31.6
	2045	19.9	9.9	4.0	1.1	34.9
	2795	24.4	10.1	2.4	1.3	38.2
	3950	25.9	7.2	5.6	3.3	42.0
	5850	34.4	12.0	5.1	1.6	53.1
	9010	51.4	17.8	8.0	2.9	80.1
average;	3765	26.4	10.7	4.9	1.8	43.7
Leucaena leucocephala;						
plants ha^{-1}	1180	23.7	24.0	13.6	6.0	67.3
	1535	30.9	22.9	9.8	2.9	66.5
	2045	32.8	44.1	12.9	5.6	95.4
	2795	30.8	34.8	19.3	7.3	92.2
	3950	27.1	41.8	14.7	7.3	90.9
	5850	47.3	42.8	18.6	5.4	114.1
	9010	60.0	35.3	28.9	7.6	131.8
average;	3765	36.1	35.1	16.8	6.0	94.0

Table 3. Fine root distribution in the three leguminous perennials (based on data from Table 2)

Species	Average plant density	0–10	10–20	20–30	30–40 cm
G. sepium	3450 pl/ha	40.8	30.3	17.8	11.1%
F. macrophylla	3765	60.3	24.4	11.2	4.1%
L. leucocephala	3765	38.4	37.3	17.9	6.4%

distribution is presented, rather than the total rootmass; Table 2. This choice is based on the fact that the fine roots are mostly responsible for water and nutrient uptake, and are as such a relevant measure for the relative competitive power of a species [12]. In the context of this discussion two aspects are important; i) the distribution of the roots in the soil profile, and ii) the weight of the rootmass.

Table 3 shows that, given the soil *stratum* sampled, about three-quarters of the fine root dry weight is concentrated in the first 20 cm of the soil profile. This is the soil layer that is also explored by the yam roots; Figure 3.

This outcome confirms the observation of Jonsson et al. (1988) that trees known to root deeply, *Leucaena leucocephala* in particular, should not be considered as having no roots in the upper soil layers. Certainly in poor soils trees are likely to root where the scarce nutrient resources are located. Consequently, crop and support tree compete for water and soil nutrients in the same layer of the topsoil.

Figure 4 shows that the total fine root dry weight per unit area of both *Flemingia macrophylla* and *Leucaena leucocephala* increases when the planting density increases, the linear fittings being highly significant in both cases. With the latter species the fine root dry weight is about twice as high.

Increasing the plant density in *Gliricidia sepium* does not result in proportionally more fine roots in the soil layer up to 40 cm depth. The fine root dry weight levels off at about 60 g m^{-2}. The much weaker developed root system in *Gliricidia sepium*, particularly compared to that of *Leucaena leucocephala*, must be considered as one of the possible factors explaining the higher tuber yield when this particular tree is used as live support in yam crop cultivation; Table 1.

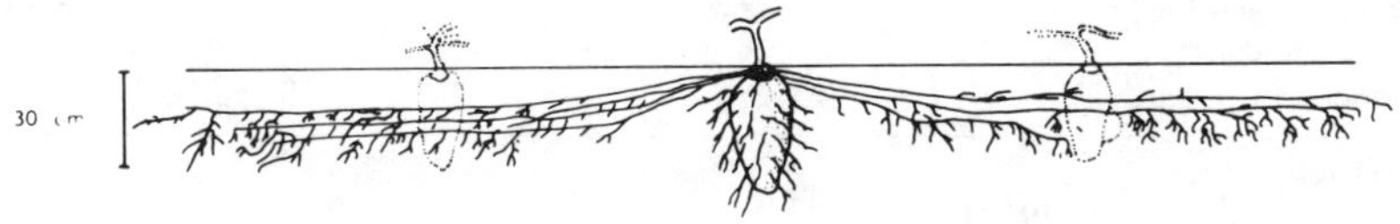

Fig. 3 . Rooting pattern of *Dioscorea alata* cv. Brazo Fuerte.

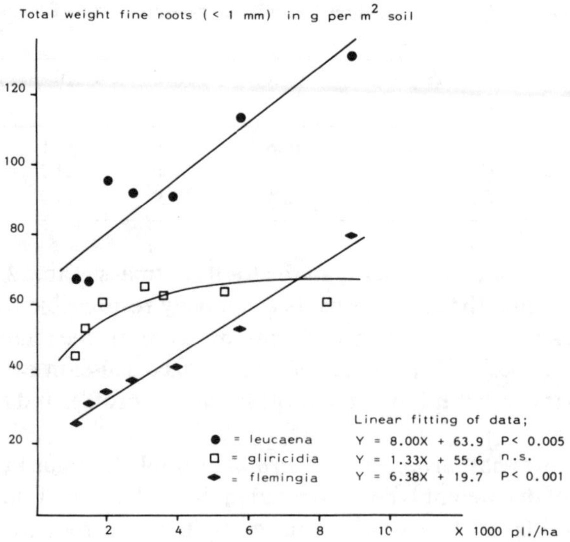

Fig. 4. Total fine root dry weight over 0–40 cm of the soil as a function of the plant density.

In Table 4 the fine root dry weight (< 1 mm) of the tree woody legumes is compared to that of several cropping systems.

Table 4 puts the differences in fine root dry weight between the woody legumes studied into perspective: The values for *Gliricidia sepium*, even at

Table 4. The fine root (< 1 mm) dry weight in the topsoil (0–25 cm) under various cropping systems

Species/plant community	g fine roots per m² soil	Planting arrangement plants per ha
– *Flemingia macrophylla*[a]	23.6–73.2*	1180–9010
– *Gliricidia sepium*[a]	38.7–47.8*	1080–8260
– *Gmelina arborea*[b]	54.2	1667;5000
– *Leucaena leucocephala*[a]	54.5–109.8*	1180–9010
– shaded coffee + *Erythrina poeppigiana*[b]	64.4	4348 + 287
– cacao + plantain + *Cordia alliodora*[b]	79.3	1111 + 1111 + 278
– wooded farm garden[b]	90.7	38 species

[a] this study
[b] data from Ewel *et al.* 1982
* interpolated from Table 2

high planting densities, are found at the lower end of the range presented, whereas those for *Leucaena leucocephala* belong to the upper part of the range.

4.4 Leaf productivity

A major difference is found in the species' relative capacity for leaf productivity. At e.g. 10,000 plants per hectare *Leucaena leucocephala* produces about 1.5 times the leaf DM *Gliricidia sepium* can produce [4]. Moreover, *Leucaena leucocephala* leaves have a higher surface (:) weight ratio compared to *Gliricidia sepium* which means that the differences in occupation of space is even larger; *Leucaena leucocephala* has a LAI 1.6 times that of *Gliricidia sepium*. Since both species attain about equal height in the cropping system studied, it can be assumed that *Leucaena leucocephala* intercepts more of the incoming light than *Gliricidia sepium*. In other words, relatively less light will be available to yam production, when the crop is grown in close association with *Leucaena leucocephala*.

At the planting density mentioned *Flemingia macrophylla* takes an intermediate position [4].

5. Conclusions

Based on the data presented, of the three leguminous perennials studied in a role as live support for yam cultivation *Leucaena leucocephala* is the least suitable. The species shows the strongest competitive power expressed in terms of leaf productivity and relative density of the rootmass in the upper soil *stratum*, the zone mostly explored by the yam crop. Indirectly, this is illustrated in the low tuber yield when yam is grown in association with *Leucaena leucocephala*.

Flemingia macrophylla is unsuitable mainly because of its structural weakness: Its branches do not sufficiently lignify to carry the yam leafmass. Moreover, its architecture is too compressed to leave sufficient space for the yam plant to develop.

The significantly higher tuber yield of yams grown with *Gliricidia sepium* is a function of specific properties of the tree species: A relatively low leaf productivity and a weakly developed root system are matched by an open architecture, which leaves sufficient space for the yam crop to develop its photosynthetic apparatus.

6. Acknowledgements

Thanks are due to Messrs E Dekker and G Limberg who undertook the yam- and tree species root studies respectively.

7. References

1. Agboola AA, Wilson GF, Getahun A and Yamoah CF (1982) Gliricidia sepium: A possible means to sustained cropping in Western Nigeria. In: MacDonald LH (Ed); Agroforestry in the African humid tropics. Tokyo, Japan. p 141–143
2. Aken'ova ME and Atta-Krah AN (1986) Control of Speargrass in an Alley-cropping fallow. NFTA Research Reports 4: 27–28
3. Budelman A (1988) The performance of the leaf mulches of Leucaena leucocephala, Flemingia macrophylla and Gliricidia sepium in weed control. Agroforestry Systems 6: 137–145
4. Budelman A (1988) Leaf dry matter productivity of three selected perennial leguminous species in humid tropical Ivory Coast. Agroforestry Systems 7: 47–62
5. Budelman A (1988) The decomposition of the leaf mulches of Leucaena leucocephala, Gliricidia sepium and Flemingia macrophylla under humid tropical conditions. Agroforestry Systems 7: 33–45
6. Budelman A (1989) Nutrient composition of the leaf biomass of three selected woody leguminous species. Agroforestry Systems 8: 39–51
7. Budelman A (1989) The performance of selected leaf mulches in temperature reduction and moisture conservation in the upper soil stratum. Agroforestry Systems 8: 53–66
8. Budowski G, Kass DCL and Russo RO (1984) Leguminous trees for shade. Pesquisa Agropecuaria Brasileira 19 s/n: 205–222
9. Clayton WD (1958) Secondary vegetation and the transition to savanna near Ibadan, Nigeria. Journal of Ecology 46 (2): 217–238
10. Ewel J, Benedict F, Berish C, Brown B, Gliessmann S, Amador M, Bermudez R, Martinez A, Miranda R and Price N (1982) Leaf area, light transmission, roots and leaf damage in nine tropical plant communities. Agro-ecosystems 7: 305–326
11. Huxley PA (1985) The tree-crop interface or simplifying the biological environmental study of mixed cropping agroforestry systems. Agroforestry Systems 3: 252–266
12. Jonsson K, Fidjeland L, Maghembe JA and Hogberg P (1988) The vertical distribution of fine roots of five tree species and maize in Morogoro, Tanzania. Agroforestry Systems 6: 63–69
13. Kang BT, Wilson GF and Lawson TL (1984) Alley-cropping; A stable alternative to shifting cultivation. International Institute of Tropical Agriculture, Ibadan. Nigeria, 23 pp
14. Onwueme IC (1978) The tropical tuber crops. Yams, Cassava, Sweet potato, and Cocoyams. John Wiley & Sons. Chichester. Section A. pp. 1–106
15. Onwueme (1981) Strategies for progress in yam research in Africa. In: Terry ER, Oduro KA and Caveness F (Ed) Tropical Root Crops Research Strategies for the 1980's. Proceedings 1st Triennial Root Crops Symposium of the International Society for Tropical Root Crops- Africa branch, 8–12 september 1980. Ibadan, Nigeria. p. 173–176
16. Wilson GF and Akapa K (1981) Improving the in-situ stem support system for yams. In: Terry ER, Oduro KA and Caveness F (Ed) Tropical Root Crops Research Strategies for the 1980's. Proceedings 1st Triennial Root Crops Symposium of the International Society for Tropical Root Crops- Africa branch, 8–12 september 1980. Ibadan, Nigeria. p 195–197

Agroforestry Systems 10: 61–69, 1990.

9

Woody legumes as live support systems in yam cultivation
*II. The Yam-*Gliricidia sepium *association*

ARNOUD BUDELMAN
Centre Néerlandais, Abidjan, Côte d'Ivoire. Department of Tropical Crop Science, Agricultural University Wageningen, The Netherlands
Present address: Royal Tropical Institute, Mauritskade 63, 1092 AD, Amsterdam, The Netherlands

Key words: *Gliricidia sepium*, Water Yam (*Dioscorea alata*), live support systems, leaf mulch

Abstract. This paper describes the effect of *Gliricidia sepium*, used as live support system on the development and tuber yield of Water Yam (*Dioscorea alata*). The leguminous perennial was planted in rows 3 m apart. The in-row distance between the trees was 1 m. In the alleys 2 rows of yam were planted, 1 m apart, with an in-row distance of 0.5 m. The aim of the experiment was to separate the effect of the leaf mulch produced by the trees, which was used to green-manure the yam crop, and the effect on the tuber yield that stems from the presence of the live stakes.

Using *Gliricidia sepium* trees as live support more than doubles the yield per unit of area of yam; 8.8 ton ha^{-1}(yam sole-cropped, control) against 20.7 ton ha^{-1}(yam-tree association). Staking explains slightly more of the yield increment than the mulch factor.

The trees and the mulch derived create more favourable conditions for development of the yam plant, while the capability of the trees to extract nutrients from soil layers that can not be reached by the yam roots is demonstrated.

It is proposed to select and breed yam cultivars directly in relation to suitable live support systems. Such integrated tree-annual crop systems potentially provide substitutes for the practice of shifting cultivation.

1. Introduction

In the first part of this paper three leguminous perennials, *Leucaena leucocephala, Gliricidia sepium* and *Flemingia macrophylla,* were compared in a role as 'in situ' grown live support systems in yam cultivation [2]. The presence of support (stakes, trellisses. *etcetera*) is known to considerably enhance yam tuber production [3], which seems to arise mainly from reducing mutual leaf shading in the yam plant [5]. In case of *live* support systems, the degree of competition excercised by the auxiliary perennial species used in that role should be minimal to allow the yam plant an adequate share of water, light and nutrients.

Of the three species in observation *Gliricidia sepium* clearly provides the most favourable environment for the yam plant, because of its low leaf productivity, its open canopy and its weakly developed root system [2].

Based on these experiences *Gliricidia sepium* was selected for further study. The objective of the experiment discussed here was to separate the contribution to the yam tuber yield that stems from the presence of stakes, from that of the application of leaf mulch produced by the trees, and, more in general, to study the dynamics of the crop-tree association.

2. Materials and methods *

The experiment, involving *Gliricidia sepium* grown in association with Water Yam (*Dioscorea alata*, cv. Brazo fuerte) included 4 treatments and 4 replications.

The treatments were:

a) control, sole-cropped yam (at a rate of 20,000 plants ha^{-1}),
b) sole-cropped yam (20,000 plants ha^{-1}), green-manured with the mulch produced in the experimental units that belong to treatment 'c',
c) yam (13,300 plants ha^{-1}) grown with *Gliricidia sepium* trees (3,300 plants ha^{-1}), while the mulch produced in situ was removed, and used in treatment 'b',
d) yam (13,300 plants ha^{-1}) cropped in association with *Gliricidia sepium* trees (3,300 plants ha^{-1}), whereas the mulch produced 'in situ' was used to green-manure the yam crop.

For the treatments 'b' and 'd' the leaf mulch application rate was 4.8 ton ha^{-1} DM. At the start of the trial the trees were 13 months old. The hedgerows of *Gliricidia sepium* were 14 m long, with an in-row spacing between the trees of 1 m. The hedgerows stood 3 m apart, while in each alley 2 rows of yam plants were planted at an in-row spacing of 0.5 m. Each experimental unit including the trees contained three rows of the woody perennial, and 4 rows of yams, with a net total of 108 yam plants to harvest. In the experimental units where yam was sole-cropped (treatments 'a' and 'b') the net total of harvestable yam plants was 135.

Soil cover by the yam crop was estimated 9 weeks after planting by taking two photographs from a fixed position (location, height, and angle) in each of the experimental units. On the photograph the area covered by the developing yam crop was copied on transparent sheets with the help of a dark-colored feltpen. A planimeter (model 'Milliplan' of METRAPLAN

*) For details on location, climate and soils see Part 1 [2].

S.A.) was subsequently used to measure the relative surfaces occupied by the yam biomass on the sheets. In statistical analysis the actual surfaces in unit of area were used. The figures were then converted into percentages soil cover.

Crop emergence and sprout length was noted 3 weeks after planting. At mid-term in the crop cycle, mature yam leaves were sampled and analysed on nitrogen and potassium to determine the crop nutrient status as a function of the different treatments. Yam leaf samples were analysed at the Central Laboratory of ORSTOM at Adiopodoumé, Côte d'Ivoire.

After c. 9 months the yam crop was harvested and tuber yields were determined at a per plant base. These figures were subsequently used to calculate yields per unit of area.

The Least Significant Difference values in Table 1 are calculated according to the *Studentised Range Test* (TUKEY's L.S.D. test), with the exception of the comparison 'stakes occupied' where STUDENT's t-test is used. Statistical significance is based on $P = 0.05$.

3. Results and discussion

Experimental data

In Table 1 observations on the development of the yam crop as a function of the different treatments are quantified. The values in each row with the same superscript are not significantly different.

Yam crop development

The combined effect of the presence of trees and the application of leaf mulch on the germination rate and sprout development of yam is significant. The data confirm the findings reported earlier as far as it concerns the influence of leaf mulch on the early development of yams setts [1]. In addition to that conclusion, Table 1 shows the influence of the trees in improving the growth conditions for yam; compare treatment 'a' (control) and treatment 'c' (live stakes, mulch removed). The positive effect of the trees on yam sett germination must be attributed to shading, which lowers temperature and limits drying out of the soil. Apparently, the shade is light enough not to interfere seriously with crop development. This is illustrated by the data on soil cover development, collected 9 weeks after planting the yam setts. At that moment of observation in the crop cycle, treatment 'd'

Table 1. Data summary; the *Gliricidia sepium*–yam association

Treatment	'a' Control	'b' Mulch, no stakes	'c' Live stakes, no mulch	'd' Live stakes + mulch	*L.S.D.* value
-yam plants ha^{-1}	20,000	20,000	13,300	13,300	
-trees ha^{-1}			3,300	3,300	
-nutrient transport	neutral	addition	extraction	neutral, recycling	
Early development					
-plants sprouted	42.5^{a}	62.7^{ab}	68.2^{b}	83.0^{b}	24.2
-sprout length cm	21.6^{a}	31.3^{bc}	26.8^{ab}	36.6^{c}	8.3
-soil cover %	36.0^{ab}	44.4^{b}	25.6^{a}	61.6^{c}	16.4
-stakes occupied by yam vines*			9.0^{a}	18.0^{b}	5.4
Nutrient content yam leaves					
-nitrogen %	3.13^{a}	3.23^{a}	3.01^{a}	3.22^{a}	*N.S.***
-potassium %	1.02^{a}	1.93^{b}	0.80^{a}	1.76^{b}	0.68
Tuber yield					
-kg per yam plant	0.44^{a}	0.89^{b}	0.54^{a}	1.55^{c}	0.29
-ton per hectare	8.8^{a}	17.8^{b}	7.2^{a}	20.7^{b}	5.1

*out of 26 live stakes sampled per experimental unit
**not significant

(live stakes + mulch) shows a significantly higher percentage of soil cover compared to all other treatments.

Soil cover in treatment 'c' (live stakes, mulch removed) is worst of all treatments, which indicates the importance of the leaf mulch as source of nutrients, whether produced 'in situ' or outside the experimental plots considered. The poor crop development in treatment 'c' (live stakes, mulch removed) is further demonstrated in the lower number of stakes occupied by yam vines, as compared to treatment 'd' (live stakes + mulch).

Yam leaf nutrient status

As a result of the application of *Gliricidia sepium* leaf mulch, the leaf nutrient status of the yam crop differs; Table 1. Nitrogen and potassium were chosen as relevant 'indicator'-nutrients. Particularly the latter is relatively scarce in the soil where the experiment took place [7]. Although the differences in nitrogen content in the yam leaves as a function of the treatments are not significant, the values reflect almost exactly the pattern found for potassium.

The potassium content in the yam leaves in the treatment 'd' (live stakes + mulch) differs significantly from treatment 'a' (control), whereas both treatments only differ in the presence of the trees. No additions, improving the soil fertility status took place.

When the mulch produced is removed from the experimental units, the potassium content in the yam leaves drops compared to the nutrient content found in the control leaf samples. The difference, 0.22%, is not statistically significant, but indicates the limited extent of the potassium reserves available. A similar observation can be made for nitrogen.

The results indicate that the trees can extract potassium from soil layers

Fig. 1. The Gliricidia-Yam association in an alley-cropping configuration (in the centre a live stake occupied by a yam vine, at the right several unoccupied live stakes).

that can not be reached by the shallow-rooting yam crop [2], making the scarce nutrient available for crop production. The capability of the tree root systems to trap nutrients at deeper soil layers and to recycle them through their biomass (mainly the leaves) is an important principle in agroforestry practices and development [8].

Quantification of cropping system components

Table 1 shows that the combined influence of live stakes and mulch application on yam tuber production is considerable. Per unit of area the yield increases by a factor 2.3, as compared to the control.

Based on the yam tuber yields per unit of area (see Table 1, last row) the relative contribution to the yam tuber yield of the two 'products' of the *Gliricidia sepium* hedges (mulch and stakes) is tentatively separated. Assumptions;

1. When yam is sole-cropped and untreated the tuber yield is noted as Ys. In the cropping systems that include one row of trees alternated by two rows of yam plants, the basic level of the tuber yield is 2/3 that of the yield of the control, hence 0.67 Ys.
2. The yam tuber yield increment explained by the mulch applied is noted as Ym. The gain from mulch application in terms of tuber yield (see eq. 2 and eq. 4; Table 2) is not necessarily equal to the loss in tuber yield when the mulch is removed (eq. 3). There are as well reasons to asssume a difference in effect of mulch applied in the two relevant cropping systems since mulch is produced outside the cropping system in one case (eq. 2) and 'in situ' in the other (eq. 4).
3. The part of the tuber yield ascribed to the 'tree-yam' interaction, $Y(t * y)$, includes the combined negative ('competition') and positive ('provision of support') influences that result from the presence of the

Table 2. Quantification of cropping system components; separating the effect of mulching and live stakes on yam tuber yields

Cropping system	Yield component equations[1]	Eq. no
–Yam sole-cropped	$Ys = 8.8$ ton	eq.1
–Yam sole-cropped, mulch added	$Ys + Ym_1 = 17.8$	eq.2
–Yam + live stakes, mulch removed	$0.67\,Ys - Ym_2 + Y(y * t)_1 = 7.2$	eq.3
–Yam + live stakes, mulch used 'in situ'	$0.67\,Ys + Ym_3 + Y(y * t)_2 = 20.7$	eq.4

[1] Data from Table 1, last row

trees. As in assumption 2, the different cropping circumstances supposedly result in different values for the interactive component; $Y(t * y)_1 \neq Y(t * y)_2$.

The equations representing 4 cropping systems include 6 parameters, which cannot be quantified unequivocally; Table 2. One solution to the problem would be to reason towards estimates.

Eq. 1 and eq. 2 can be used to calculate values for the control yield level and the effect of mulch application when yam is sole-cropped; $Ys = 8.8$ and $Ym_1 = 9.0$ ton ha^{-1}. The yield component explained as a function of the application of the mulch (4.8 ton ha^{-1} DM leaf mulch) is 9.0 ton, or c. 1.9 ton tubers per ton leaf mulch DM of *Gliricidia sepium*. This ratio is realistic and has been found in another trial [1].

In the cropping system represented by eq. 4 a similar quantity of mulch was applied (at a rate of 4.8 ton ha^{-1} DM) and distributed evenly in the experimental units. It would, however, be incorrect to assume a yield improvement of 9.0 ton tubers as a result of the mulch application. Firstly, only 2/3 of the yam plants are available, compared to sole-cropping, that can profit from the nutrients released. Secondly, it is very likely that part of the nutrients from the mulch is directly re-used by the trees present in the cropping system, hence $Ym_3 < Ym_1$. If we assume that the trees have equal chance to absorb the nutrients released compared to the neighbouring yam plants $Ym_3 = 2/3\,Ym_1$.

Eq. 4 then can be solved; mulching explains 6.0 tons, and staking 8.8 tons of the total yield.

Similar reasoning can be applied to solve eq. 3. Extraction of nutrients by the trees at least partly takes place outside the rooting space occupied by the yam plants, so the yam crop will not suffer from the full impact of the nutrient removal from the soil; $Ym_2 < Ym_1$. Assuming $Ym_2 = 2/3\,Ym_1$, the nutrients extracted in cropping system 3 result in a potential loss of 6.0 ton yam tubers, and a gain from the presence of the live stakes, $Y(y * t)_1$, of 7.3 ton of yam tubers.

It can be concluded that the live stakes explain a slightly larger part of the tuber yield increment than the mulch produced by the trees. Over a range of values for $Ym_{2,3}$ of 5 to 9 ton tubers, $Y(y * t)_1$ always $> Ym_2$ and $Y(y * t)_2 > Ym_3$ as long as $Ym_3 < 7.4$ ton.

5. Conclusions

The results show that *Gliricidia sepium* is a suitable species to act as 'in situ' support system for yam cultivation, because of the limited competition the tree exercises [2], as well as the beneficial, complementary effects described in this paper. The light, regulated shade from the trees in combination with proper application of the leaf mulch produced improve the micro-climate for the yam plant, reducing the possibility of stress. Secondly, the mulch represents a concentration of nutrients, a product that can be used to enrich the soil where it matters most, that is close to the yam roots.

Apart from improving the productivity of yam, live support systems represent an important additional value that is not found when yams are sole-cropped, since the trees can function as a planted fallow during off-season. This makes the tree-based cropping system version of yam cultivation a potentially interesting start in the search of substitutes for the practice of shifting cultivation.

Obviously, the results presented here are only preliminary and do not warrant suggestions regarding the long term viability of a yam-based cropping system including *Gliricidia sepium* trees. Follow-up research is needed, which must include aspects such as crop rotation and fertiliser use. An intriguing question, for example, will be whether the tree-mulch system will enhance the efficiency of fertiliser use.

The results are possibly also of interest in the field of selection and breeding of yam cultivars. Onwueme's suggestion to breed yam cv's that do well without staking and to mechanise its cultivation apparently is not the only solution to declining yields in areas where stakes are difficult to obtain [6]. Yam plants possess an indeterminate growth habit and profit from a support structure in terms of an increased photosynthetic apparatus leading to higher tuber yields. This natural inclination finds a better expression when support systems are provided.

An alternative approach to the option proposed by Onwueme therefore would be to select *directly* in an intercropping environment, and to consider live support systems as an integral part of the breeding aproach.

Within the context of this discussion, climbing beans intercropped with maize provide interesting parallel evidence. Zimmermann *et al.* [1984] show that the heritability of yield for beans in early generations of a breeding programme is higher in an intercropping situation with maize than when sole-cropped. The explanation offered by Davis *et al.* [1986] is that the competitive ability of the bean plant has a higher heritability than yield potential, because this aspect is related with highly heritable characteristics such as plant height and vigour. Simply stated; give the climbing bean plant the opportunity to fully develop and good yields will follow as a consequence.

This reaction is probably also valid in the case of *Dioscorea* spp., since

Table 1 clearly illustrates the significant interaction between the growth habit of a crop and growth circumstances.

This evidence justifies breeding with a cropping system perspective.

References

1. Budelman A (1989) Effect of the application of the leaf mulch of *Gliricidia sepium* on early development, leaf nutrient content and tuber yields of Water Yam (*Dioscorea alata*). Agroforestry Systems 8: 243–256
2. Budelman A (in press) Woody legumes as live support systems in yam cultivation. I. The tree-crop interface. Agroforestry Systems
3. Campbell JS and Gooding HJ (1962) Recent developments in the production of food crops in Trinidad. Tropical Agriculture (Trin.) 30: 261–270
4. Davis JHC, Woolley JN and Moreno RA (1986) Multiple cropping with legumes and starchy roots. In; Francis CA (Ed.) Multiple cropping systems. pp 133–160. Macmillan, NY and Collier Macmillan, London
5. Onwueme IC (1978) The tropical tuber crops. Yams, Cassava, Sweet potato, and Cocoyams. John Wiley & Sons. Chicester. Section A, pp 1–106. ISBN O 471 99607 4
6. Onwueme IC (1981) Strategies for progress in yam research in Africa. In: Terry ER, Oduro KA and Caveness F (ed.) Tropical Root Crops Research Strategies for the 1980's. Proceedings 1st Triennial Root Crops Symposium of the International Society for Tropical Root Crops — Africa branch, 8–12 September 1980. pp 173–176. Ibadan, Nigeria
7. Roose E and Chéroux M (1966) Les sol du bassin sédimentaire de Côte d'Ivoire. Cahier ORSTOM-Série Pédologie 4(2): 51–92
8. Young A (1989) Ten hypotheses for soil-agroforestry research. Agroforestry Today 1(1): 13–16
9. Zimmermann MJO, Rosielle AA and Waines JG (1984) Heritabilities of grain yield of common bean in sole crop and in intercrop with maize. Crop science 24: 641–644